I0482697

Disclaimer

Book Title: Heating Mode Performance Measurements For A Residential Heat Pump With Single-Faults Imposed

Book Author: William V. Payne; Piotr A. Domanski; Seok H. Yoon;

Book Abstract: The objective of this research was to acquire heating mode experimental data for a residential heat pump needed for the development of a FDD methodology. Within this project, a heat pump equipped with a thermostatic expansion valve (TXV) was tested in the NIST environmental chambers during steady-state no-fault and imposed-fault operation. Six faults were imposed: 1) compressor valve leakage, 2) outdoor improper airflow, 3) indoor improper airflow, 4) liquid line restriction, 5) refrigerant undercharge, and 6) refrigerant overcharge. Transient tests were also performed to acquire data necessary for the development of a FDD steady-state detector.

Citation: NIST TN - 1648

Keywords: fault detection and diagnosis; heating mode; heat pump; vapor compression

NIST Technical Note XXXX

HEATING MODE PERFORMANCE MEASUREMENTS FOR A RESIDENTIAL HEAT PUMP WITH SINGLE-FAULTS IMPOSED

W. Vance Payne
Seok Ho Yoon
Piotr A. Domanski

NIST National Institute of Standards and Technology • U.S. Department of Commerce

NIST Technical Note XXXX

HEATING MODE PERFORMANCE MEASUREMENTS FOR A RESIDENTIAL HEAT PUMP WITH SINGLE-FAULTS IMPOSED

W. Vance Payne
Piotr A. Domanski
U.S. DEPARTMENT OF COMMERCE
National Institute of Standards and Technology
Building Environment Division
Building and Fire Research Laboratory
Gaithersburg, Maryland 20899-8631

Seok Ho Yoon
Energy Systems Research Division
Korea Institute of Machinery and Materials
171 Jang-dong, Yuseong-gu, Daejeon 305-343, Korea

August 2009

U.S. DEPARTMENT OF COMMERCE
Gary Locke, Secretary

NATIONAL INSTITUTE OF STANDARDS AND TECHNOLOGY
Patrick D. Gallagher, Deputy Director

Certain commercial entities, equipment, or materials may be identified in this document in order to describe an experimental procedure or concept adequately. Such identification is not intended to imply recommendation or endorsement by the National Institute of Standards and Technology, nor is it intended to imply that the entities, materials, or equipment are necessarily the best available for the purpose.

National Institute of Standards and Technology Technical Note XXXX
Natl. Inst. Stand. Technol. Spec. Publ. XXXX, NNN pages (August 2009)
CODEN: NSPUE2

TABLE OF CONTENTS

LIST OF TABLES

LIST OF FIGURES

NOMENCLATURE

Symbols

m_R	refrigerant mass flow rate, kg min^{-1}
P	pressure, kPa absolute
$P_{TXV,up}$	upstream pressure of TXV, kPa
ΔP_{CR}	condenser refrigerant side pressure drop, Pa
ΔP_{ER}	evaporator refrigerant side pressure drop, Pa
ΔP_{LL}	liquid line pressure drop, Pa
Q	capacity, W
Q_{CA}	condenser air-side capacity, W (including fan heat)
Q_{CR}	condenser refrigerant-side capacity, W (adjusted for fan heat)
Q_{EA}	evaporator air-side capacity, W (includes fan heat)
Q_{ER}	evaporator refrigerant-side capacity, W (adjusted for fan heat)
T	temperature, °C
T_{CR}	condenser inlet saturation temperature, °C
T_{C15}	indoor coil return bend thermocouple TC#15, °C
T_D	compressor discharge-line wall temperature, °C
T_{ER}	evaporator exit saturation temperature, °C
T_{E103}	evaporator bend thermocouple, TC#103, °C
T_{ID}	indoor dry-bulb temperature, °C
T_{IDF}	indoor fan case temperature, °C
T_{IDP}	indoor dew-point temperature, °C
T_{OD}	outdoor dry-bulb temperature, °C
T_{ODF}	outdoor fan case temperature, °C
T_{ODP}	outdoor dew-point temperature, °C
T_{SW}	compressor suction-line wall temperature, °C
ΔT_{CA}	condenser air temperature rise, °C
ΔT_{EA}	evaporator air temperature drop, °C
ΔT_{LL}	liquid line temperature drop, °C
ΔT_{RVD}	reversing valve temperature change, discharge side, °C
ΔT_{RVS}	reversing valve temperature change, suction side, °C
ΔT_{sc}	liquid line subcooling, °C
ΔT_{scV}	liquid line subcooling at outdoor service valve, °C
ΔT_{shC}	condenser inlet superheat, °C
ΔT_{shE}	evaporator exit superheat, °C
ΔT_{shV}	vapor superheat at outdoor service valve, °C
ΔT_{103}	outdoor temperature minus TC#103, °C
W_{comp}	compressor work, W

Abbreviations

AC	air conditioner (conditioning)
CF	condenser fouling (improper outdoor airflow)
CMF	compressor valve or 4-way reversing valve leakage fault
COP	Coefficient of Performance (−)
DB	dry bulb
DOE	U.S. Department of Energy
EF	evaporator fouling (improper indoor airflow)
EPA	U.S. Environmental Protection Agency
FDD	fault detection and diagnosis (diagnostics)

HP	heat pump
HVAC	heating, ventilating and air-conditioning
IC	improper charge of refrigerant
ID	indoor
LL	liquid line restriction fault
NFSS	no-fault steady-state
OC	refrigerant overcharge
OD	outdoor
RH	relative humidity
R	residual of features
SEER	Seasonal Energy Efficiency Ratio (Btu/(W h))
SHR	Sensible Heat Ratio (–)
TXV	thermostatic expansion valve
UC	refrigerant undercharge
WB	wet bulb
w/	with
w/o	without
wrt	with respect to

CHAPTER 1. Introduction

An increasing emphasis on energy saving and environmental conservation requires that air conditioners and heat pumps be highly efficient. To this end, several government initiatives have been undertaken. For example, the U.S. Environmental Protection Agency (EPA)'s Global Programs Division is responsible for the assessment of alternative refrigerant performance and enforcement of the Clean Air Act. Another prime example is the ENERGY STAR initiative, a program formulated by the EPA/Climate Protection Partnerships Division and the Department of Energy (DOE), which promotes products that offer energy efficiency gains and pollution reduction.

To assure that heating, ventilating, air-conditioning (HVAC) equipment operates in the field at its design efficiency, the efforts exerted by equipment manufacturers to improve equipment SEER must be paralleled in the field by good equipment installation and maintenance practices. However, a survey of over 55000 residential and commercial units found the refrigerant charge to be incorrect in more than 60 % of the systems (Proctor, 2004). Another independent survey of 1500 rooftop units showed that the average efficiency was only 80 % of the expected value, primarily due to improper refrigerant charge (Rossi, 2004). A low refrigerant charge in the system may be due to a refrigerant leak or improper charging during system installation. While the most common concern about a refrigerant leak is that a greenhouse gas has been released to the atmosphere, a greater impact is caused by the additional CO_2 emissions from fossil fuel power plants due to the lowered efficiency of the air-conditioning (AC) unit.

Proctor's survey (2004) shows a correlation between the quality of installation and technician training and supervision. Proper training of the technician is a necessary requirement for proper installation. But the survey also showed clearly that the number of return calls to correct improper installation was lowest when routine oversight of the installation work was provided, and that the number of faulty installations markedly increased when post-installation inspection visits were eliminated. At present, the homeowner has no mandated quality assurance method for equipment installation as long as his/her comfort is not compromised. Proctor (2002) discussed a fault diagnosis service marketed by his company that aids technicians in repairing systems; the technician phones-in equipment specifications and current readings and receives a check-list with other diagnostic and repair suggestions over the phone. The Air Conditioning Contractors of America (ACCA) has been concerned with this problem and developed ACCA Standard 5 (ANSI/ACCA Standard 5 QI-2007); their standard titled "HVAC Quality Installation Specification" seeks to remove ambiguity in system installations and promote consistency and quality among system installers.

The goal of this project is to study and develop fault detection and diagnostic (FDD) methods which would provide a technician with a fault diagnosis and could alert a homeowner when performance of their AC unit falls below the expected range, either during commissioning or post-commissioning operation. For the homeowner, this FDD capability could be incorporated into a future smart thermostat where a readout on performance would allow basic oversight of the service done on the unit and register the effects of that service upon performance.

FDD methods for ACs and HPs will contribute to;
- reduction of energy use
- reduction in peak demand of electricity
- reduction in CO_2 emissions from fossil fuel power plants
- reduction in refrigerant emissions from AC and HP systems
- reduction in down time and maintenance cost
- improvement in thermal comfort.

Arguably, development of FDD methods for split equipment presents unique challenges because these systems are assembled on site. Varied assembly skill levels and lack of attention to manufacturer recommendations is the prime reason that automated FDD methods be developed. FDD will provide "Automated Oversight" of servicing and warning of refrigerant charge loss, which is the most frequent problem in field assembled systems. It should be noted that the refrigerant leak problem may become even more frequent with the industry transition from medium pressure R22 to higher pressure R410A.

Fault detection and diagnosis is accomplished by comparing a system's current performance or parameters with those expected based on the measurements taken from the system when it was known to operate fault-free. Consequently, FDD method development includes a laboratory phase during which fault-free and faulty operations are mapped. The faults are artificially imposed to learn about the system's response to them. The analytical phase, which follows, is concerned with using the obtained database to develop methods for fault detection and diagnosis. This report documents the laboratory phase of the FDD study carried out on a residential heat pump with a Thermostatic Expansion Valve (TXV) operating in the heating mode.

Fault detection and diagnosis has been effectively applied for some time in critical systems and processes, e.g., aerospace and nuclear industry applications, and in chemical processes. FDD methods for HVAC&R systems have been under development since the late 1980's (McKellar, 1987; Pape et al., 1991; Grimmelius et al., 1995; Stylianou and Nikanpour, 1996). The majority of the early research was devoted to variable air volume air-handling units. On the vapor-compression side, most work was devoted to large systems, particularly to water chillers and single-package air conditioners (Rossi and Braun, 1997; Castro, 2002; Li and Braun, 2003), while split air-conditioning and commercial refrigeration systems received little attention.

CHAPTER 2. Literature Review

2.1 Research Background

FDD systems were originally developed as part of the fail-safe monitoring systems for the purpose of safety for nuclear power plants or aircraft (Braun, 1999). In such applications, FDD systems are equipped for fail-safe operation regardless of cost. On the other hand, a number of industrial applications pursue the reduction of total costs related to equipment downtime, service costs, and utility costs. FDD systems may be applied to reduce costs associated with all of these concerns.

As a result of the decreasing price of sensors and microprocessors, developers can affordably apply FDD systems to automatic management of even non-critical HVAC systems. In addition, remote management systems are being developed using information-based network approaches to increase energy efficiency (Snoonian, 2003). A number of research efforts for optimized management systems have been carried out in order to reduce energy consumption (Brownell et al., 1999; Seem et al., 1999; Hayter et al., 1999; Breuker et al., 2000; Roth et al., 2005)

The energy savings attributable to FDD depends on the frequency and severity of faults. A brief note based on interviews with practicing engineers and contractors reported that inefficient operation wastes at least 20 % to 30 % of the entire HVAC energy consumption (Westphalen et al., 2003). For rooftop air-conditioning units, the average efficiency was estimated at 80 % of the expected value. Approximately 50 % of installations were reported to have efficiency of 80 % or less of their design efficiency, and 20 % of installations had efficiency of 70 % or less of their design efficiency (Rossi, 2004).

Proctor (2004) surveyed over 55,000 commercial and residential air-conditioning units in California. Proctor reported that residential systems are better managed than commercial systems; however, their overall quality control was poor. From the report, 95 % of residential units failed the diagnostic test. The main reasons causing the failures are listed as duct leakage, poor insulation, excessive resistance to airflow and low evaporator airflow, improper refrigerant charge, or over-sized units. Furthermore, the refrigerant charge in residential air-conditioning units was incorrect 62 % of the time, and charge in commercial units was incorrect 60 % of the time. Proctor also provided other survey results concerning improper installation and management problems. Proctor's survey shows a correlation between the quality of installation and the technician's training and supervision. Another reference by Proctor (2002) also discusses issues related to servicing and common faults.

2.2 Previous Research

Initial FDD research in the HVAC&R field was performed for variable air volume air-handling units and chillers. During the development of the first FDD techniques, energy savings was a secondary consideration to preventing equipment malfunction. Anderson et al. (1989) studied statistical analysis preprocessors and rule based expert systems to monitor and diagnose HVAC systems. Pape et al. (1991) developed a methodology for fault detection in HVAC systems based on optimal control. In order to detect faults in system operation, deviation from optimal performance was sensed by comparing the measured system power with the power predicted using the optimal control strategy. Norford and Little (1993) presented a method for diagnosing faults in HVAC systems using parametric models of consumed electric power.

Lee et al. (1996a) represented a scheme for detecting faults in an air-handling unit using recursive parameter identification methods. One approach used in that study was to define residuals that represent the difference between the existing state of the system and the normal state. Residuals that are significantly different from zero represent the occurrence of a fault. In a successive investigation by Lee et al. (1996b), they described the application of an artificial neural network to the problem of fault diagnosis. If the system being monitored is complex, isolating the fault can be challenging, and diagnostic tools should be more adaptable for this purpose. They showed that the artificial neural network method can be a good solution to such problems.

Peitsman and Bakker (1996) applied a black-box model to an HVAC system for fault detection. Multi-input and single-output (MISO) autoregressive with exogenous input models and artificial neural network models are used in the study. The whole HVAC system is regarded as a block box instead of as a collection of component models. With the component model type, the components of the HVAC system are regarded as separate black boxes.

Only recently have investigators begun to examine FDD techniques for vapor compression systems rather than the broader area of the whole HVAC system. FDD for vapor compression systems was initially intended to help technicians servicing individual vapor compression systems. Grimmelius (1995) developed an on-line failure diagnosis system for a vapor compression refrigeration system used in a naval vessel or a refrigerated plant. He established a symptom matrix based on the combination of casual analysis, expert knowledge, and simulated failure modes. Using fuzzy logic, a real-time recognition of the failure model was suggested. The author commented on the need to develop more general skills for reference state estimation, on insensitive pattern recognition routines for failure models, and on transient diagnostic models.

Stylianou and Nikanpour (1996) represented a methodology using thermodynamic modeling, pattern recognition, and expert knowledge to determine the health of a reciprocating chiller and to diagnose selected faults. The authors suggested three fault detection modules for startup, stop, and steady-state operations based on a thermodynamic model and expert knowledge of the chiller. They tried to deal with the sensor drift fault when the chiller was off. In a successive investigation, Stylianou (1997) presented a fault diagnostic methodology using a Bayesian decision rule which assigned different faults, including no-fault, status to single classes.

Rossi and Braun (1997) developed a statistical FDD method for a roof-top air conditioner. The fault diagram was developed with temperature measurements. The residual values are used as performance indices for both fault detection and diagnosis. Statistical properties of the residuals for current and normal operation are used to classify the current operation as faulty or normal. Five kinds of faults can be distinguished from the diagnosis. They suggested a fault detection classifier and a fault diagnostic classifier. The fault detection classifier module was based on a Bayesian decision rule, and the fault diagnostic classifier module was developed assuming individual features as a series of independent probabilistic events.

Breuker and Braun (1998) surveyed frequently occurring faults for a packaged air conditioner using field data. Based on the field data, Breuker and Braun sorted field faults into three different categories according to the cause of the fault, service frequency, and service cost. With respect to the cause of the fault, system shutdown failures were caused by electrical or control problems approximately 40 % of the time and mechanical problems approximately 60 % of the time. When sorted by service frequency, refrigerant leakage occurs most frequently, followed by condenser, air handling, evaporator, and compressor faults. When sorted by service cost, compressor failure contributes 24 % of total service costs. Control related faults contribute 10 % of total service costs.

Chen and Braun (2001) developed a simplified FDD method for a 17.6 kW rooftop air conditioner with a TXV. They modified an FDD technique by simplifying Rossi and Braun's method (1997). They used measurements and model predictions of temperatures for normal system operation to compute ratios which were sensitive to individual faults. They also proposed a simple rule-based FDD process of sequential rules developed by comparing the sensitivity of residuals organized within a fault characteristic chart.

Castro (2002) applied a k-nearest neighbor and k-nearest prototype method for fault detection of a chiller. The author calculated Euclidean distances for the current state based on the selected two largest residuals, and estimated the possibility of a fault from the distance information. In this research, the software MATCH was developed as a tool for the controls package to combine monitoring, fault detection, and diagnostic features. After detecting faults, data deemed faulty were input to the rule-based fault diagnosis algorithm. Castro preferred the nearest prototype classifier since the nearest neighbor classifier is more computationally intensive.

Comstock and Braun (2001) tested eight common faults in a 316 kW centrifugal chiller to identify the sensitivity of different measurements to faults. The identification of common faults was determined through a fault survey among major American chiller manufacturers. The fault testing led to a set of generic rules for the impacts of faults on measurements that could be used for FDD. Impact of faults on cooling capacity and coefficient of performance were also evaluated.

Smith and Braun (2003) performed field-site tests on more than 20 units to identify local installation and operation problems. Using a 10.6 kW rooftop unit with a fixed orifice expansion device and a 17.6 kW unit with a TXV, the FDD problem was formulated in a mathematical way and a decoupling based unified FDD technique was proposed to handle multiple simultaneous faults.

Li (2004) re-examined the statistical rule-based method initially formulated by Rossi and Braun (1997) and presented two additional FDD schemes which improved the sensitivity of the FDD module. He also provided virtual sensors to estimate characteristic parameters from indirect component modeling. For a reference model, Li combined a multivariate polynomial model and a generalized regressive neural network (GRNN).

Kim and Kim (2005) tested a water-to-water heat pump system with a variable speed compressor and an electrical expansion valve (EEV). From the research, the system parameters are found to be less sensitive to faults compared to a constant speed compressor system. They reported that controlling the compressor speed suppressed the changeability of the system. They also provided an FDD algorithm along with two different rule-based charts depending on the compressor status. Kim and Kim suggested that COP degradation due to a fault is much more severe with a variable speed compressor than with a constant speed compressor.

Li and Braun (2007) examined a large amount of data to determine general features for any vapor compression system that clearly indicated a particular fault regardless of load level. This "decoupling" technique was shown to produce accurate indications of individual system faults even in the presence of multiple and simultaneous faults. This is a powerful technique that uses the concept of virtual sensors in vapor compression systems. Li and Braun (2009) also applied this technique to check valves and four-way refrigerant reversing valve leakage faults in the cooling mode.

Kim et al. (2006) documented the cooling mode performance of a residential heat pump equipped with a thermostatic expansion valve, TXV. The results showed that the active control of refrigerant flow by the TXV tended to mitigate certain faults and that the TXV actuation limits (fully opened or closed) had to be accounted for in developing an FDD scheme. The work was further extended to a complete FDD algorithm (Kim et al. 2008a) utilizing the techniques presented by Li (2004).

2.3 Research Objective

The objective of this research was to acquire heating mode experimental data for a residential heat pump needed for the development of a FDD methodology. Within this project, a heat pump equipped with a thermostatic expansion valve (TXV) was tested in the NIST environmental chambers during steady-state no-fault and imposed-fault operation. Six faults were imposed: 1) compressor valve leakage, 2) outdoor improper airflow, 3) indoor improper airflow, 4) liquid line restriction, 5) refrigerant undercharge, and 6) refrigerant overcharge. Transient tests were also performed to acquire data necessary for the development of a FDD steady-state detector.

CHAPTER 3. Experimental Setup and Test Procedure

3.1 System Selected for Testing

The studied system was an R410A split residential heat pump of a 8.8 kW nominal cooling capacity, Seasonal Energy Efficiency Ratio (SEER) of 13, and Heating Seasonal Performance Factor (HSPF) of 7.8 (ARI 2006). The system comprised the indoor fan-coil section, outdoor section with a compressor and four-way valve, a cooling mode and heating mode thermostatic expansion valve (TXV), and connecting tubing. Both the indoor and outdoor air-to-refrigerant heat exchangers were of the finned-tube type. The system was installed in NIST's environmental chambers and charged with refrigerant in the cooling mode according to the manufacturer's specifications. Refrigerant charge was set by noting the temperature difference between the outdoor air and the refrigerant liquid line; refrigerant was added to decrease the temperature difference to the manufacturer's specified value.

Figure 3.1 shows the outdoor section. A flow guide was attached to the top of the unit to aid in the traverse of a hot wire anemometer. The coil had 81 cm x 175 cm of finned area, with 7 fins cm^{-1} of a wavy-lanced type fins. Figure 3.2 shows a schematic of the outdoor heat exchanger refrigerant circuitry.

Figure 3.1. Side view of the outdoor section with the flow guide

Figure 3.2. Circuitry of the outdoor coil

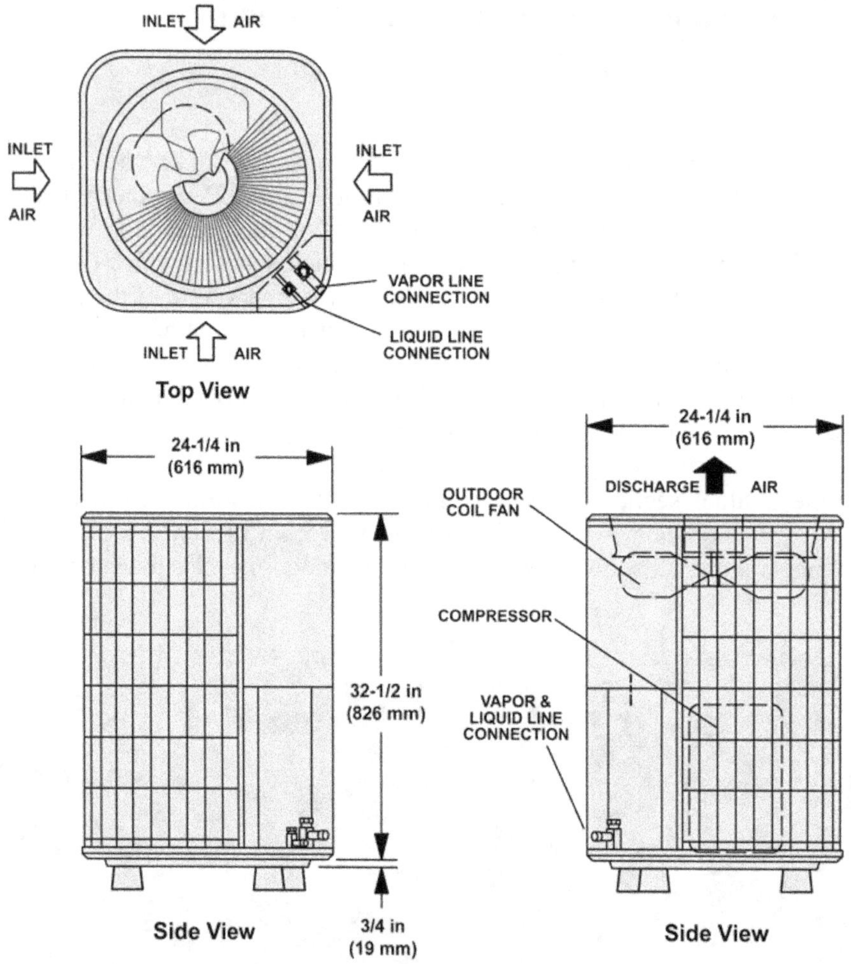

Top View

Side View   3/4 in (19 mm)   **Side View**

Figure 3.3. Outdoor section dimensions

The circles denote the tubes, the continuous lines indicate the return bends on the near side of the heat exchanger, and the broken lines indicate the return bends on the far side. As Figure 3.2 shows, the outdoor coil had five inlets and five outlets, i.e., it had five independent circuit branches. Figure 3.3 presents a graphical representation and dimensions of the outdoor unit.

Figure 3.4 shows the side view of the indoor coil assembly. It comprised two identical slabs and was designed for airflow from the bottom to the top. The two slabs were configured in an A-shape at an angle of 60°. Each slab of the coil had 48.5 cm x 43 cm of finned area with 5 fins cm^{-1} of the wavy-lanced type. Figure 3.5 shows the refrigerant circuitry for both slabs. Each slab had two inlets from the TXV and two outlets connected to the suction manifold. Figure 3.6 shows the configuration and outside dimensions of the indoor fan-coil unit. The airflow rate through the coil was approximately 1700 m^3 h^{-1} during standard cooling tests. With the same fan speed, heating airflow rate varied from 1650 m^3 h^{-1} to 1675 m^3 h^{-1}. Fin thickness for both the indoor and outdoor coils was 0.12 mm. The indoor blower and system functions were controlled through normal thermostat low voltage wiring, powered by the airhandlers low voltage transformer, connected to single-pole single-throw (SPST) manually activated switches.

Figure 3.4. Side view of the indoor coil in an upflow configuration

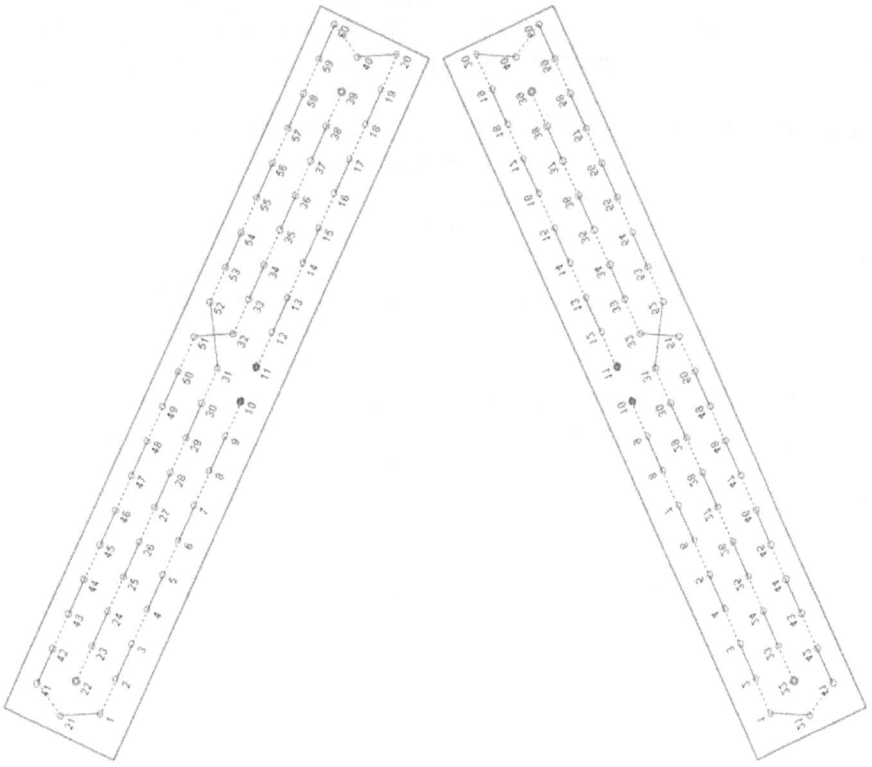

Figure 3.5. Circuitry of indoor coil

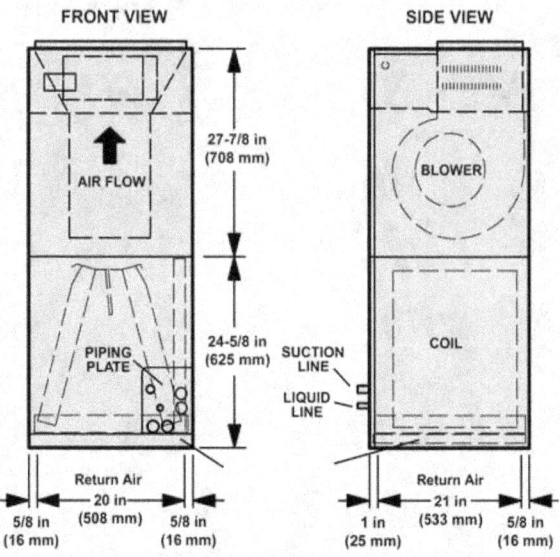

Figure 3.6. Indoor fan-coil unit dimensions

3.2 System Setup and Instrumentation

Figure 3.7 shows the air duct arrangement in the indoor environmental chamber. The ductwork was constructed according to the applicable standards (ANSI/ASHRAE Standard 51, ANSI/ASHRAE Standard 37). The air was pulled through the test apparatus by a centrifugal fan located at the outlet of the nozzle chamber ductwork. Figure 3.8 shows the schematic diagram of the heat pump installation. On the air side, the setup involved measurements of dry-bulb and dew-point temperatures, barometric pressure and pressure drop. Dew-point temperature was measured at the inlet of the evaporator ductwork and downstream of the evaporator and air mixers. Twenty-five node, T-type thermocouple grids and thermopiles measured air temperatures and temperature change, respectively.

On the refrigerant side, pressure transducers and T-type thermocouple probes were attached at the inlet and exit of every component of the system to measure the actual refrigerant pressure and temperature. Indoor coil inlet and exit temperature were measured within oil filled thermowells extending at least four inches into the refrigerant flow. Surface mounted thermocouples were secured by copper tape after being embedded in thermally conductive paste; the copper tape was further secured by a small zip-type plastic tie. The thermocouple wire was laid down on top of the tubing for several inches and further secured by zip-type ties before being well insulated with foam tape and foam insulation. The refrigerant mass flow rate was also measured using a Coriolis flow meter. The air enthalpy method served as the primary measurement of the system capacity, and the refrigerant enthalpy method served as the secondary measurement. These two measurements always agreed within 5 %. Additionally, compressor power was measured for calculations of the coefficient of performance (COP). Table 3.1 lists characteristic uncertainties of the major quantities measured during this work. A detailed uncertainty analysis may be found in Payne et al. (1999).

Table 3.1. Measurement uncertainties

Measurement	Range	Total Uncertainty at a 95 % Confidence Level
Individual Temperature	-18 °C to 93 °C	±0.3 °C
Temperature Difference (25 junction thermopile)	0 °C to 28 °C	±0.3 °C
Refrigerant Pressure	0 kPa to 3500 kPa absolute	±1.0 % of reading
Refrigerant Pressure Difference	0 kPa to 100 kPa	±1.0 kPa
Air Nozzle Pressure	0 Pa to 1245 Pa	±1.0 Pa
Refrigerant Mass Flow Rate	0 kg h^{-1} to 544 kg h^{-1}	±1.0 %
Dew-Point Temperature	0 °C to 38 °C	±0.4 °C
Dry-Bulb Temperature	1 °C to 38 °C	±0.4 °C
Total Capacity	4.3 kW to 10.6 kW	±4.0 %
COP	2.5 to 6.0	±5.5 %

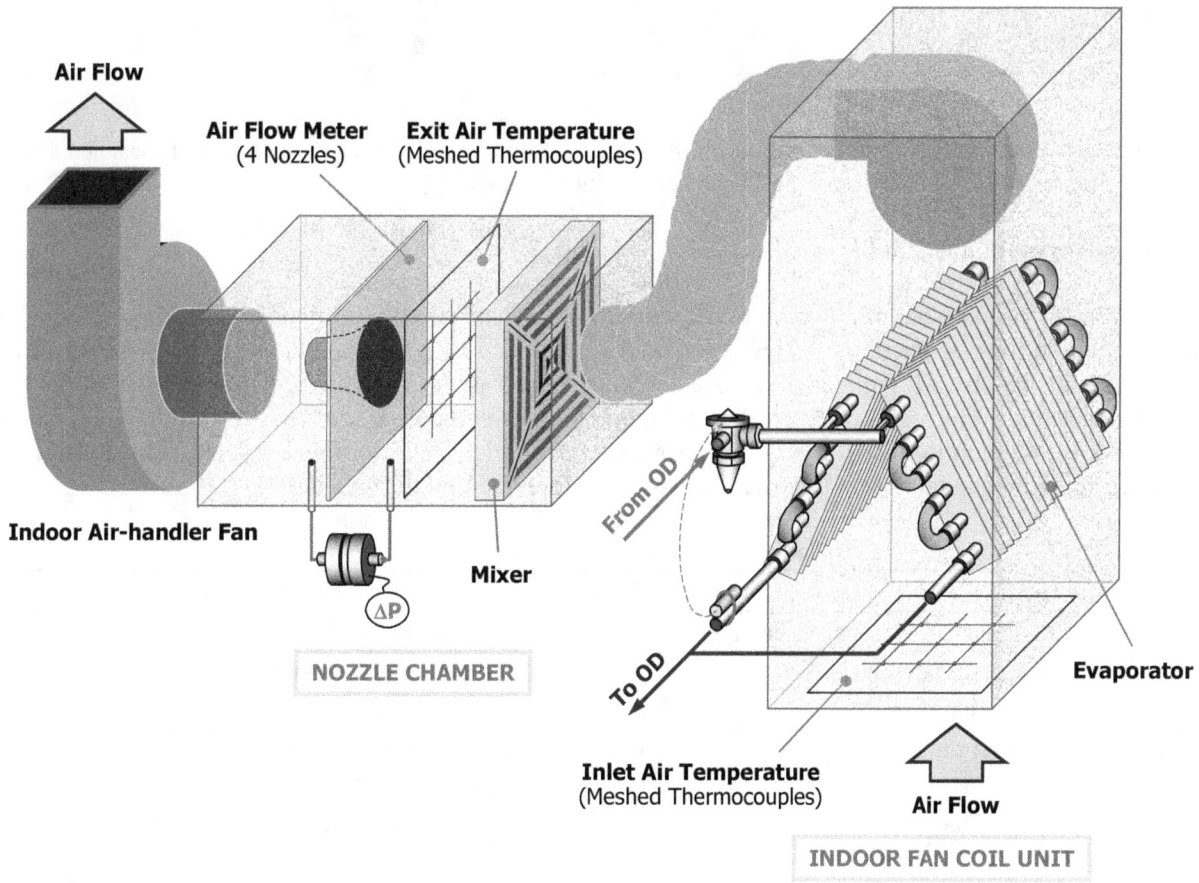

Figure 3.7. Air duct arrangement in the indoor environmental chamber

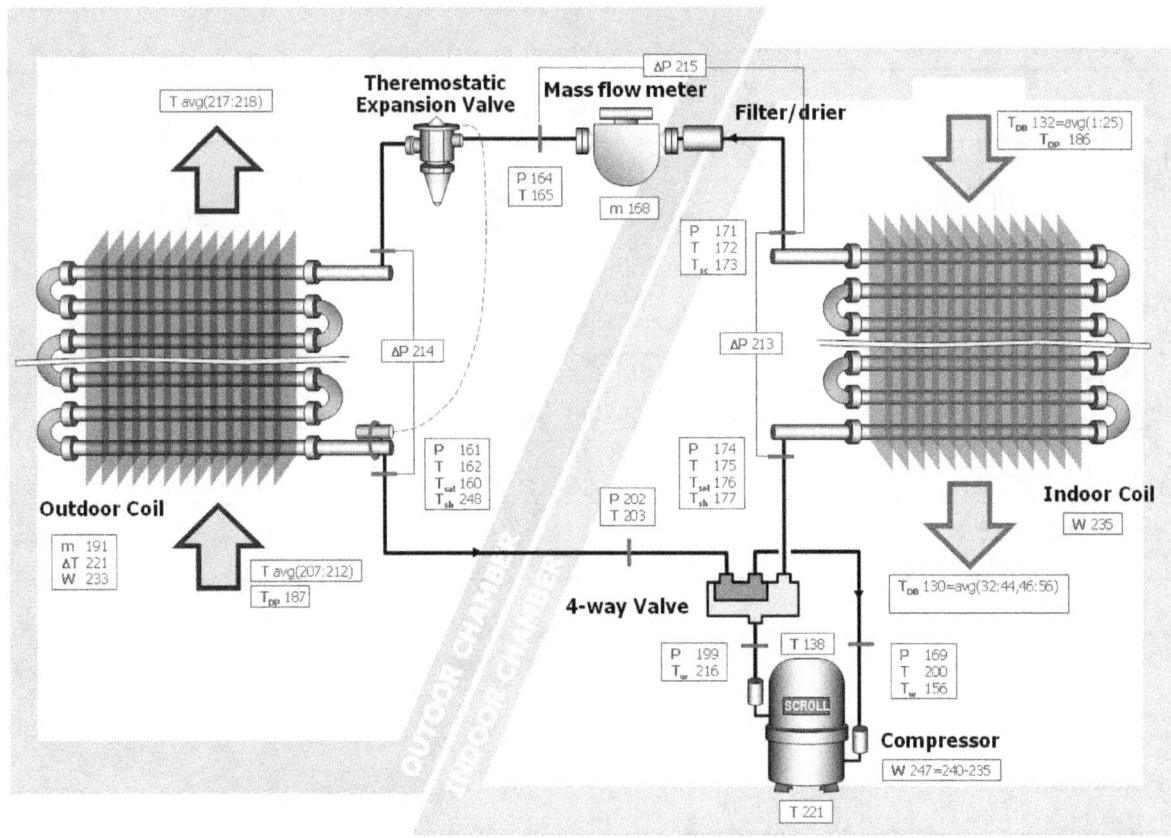

Figure 3.8. Schematic of the heat pump in the heating mode (numbers indicate references to scanned data array and are for internal use only)

3.3 Implementation of Faults

Table 3.2 lists the six types of common faults which were investigated in this study. The improper indoor airflow, liquid line restriction, refrigerant overcharge, and refrigerant undercharge may result from improper installation. All faults in Table 3.2 may also appear in the system after installation over the life of the heat pump. The respective causes may be compressor valve wear, outdoor coil fouling, a dirty air filter or coil fouling, dirty filter/dryer, improper recharge service, and a refrigerant leak. These six faults and their implementation during tests are discussed in more detail in the sections below.

Table 3.2. Description of studied faults

Fault	Abbr.	Determination of level of fault during tests
Compressor leakage (4-way valve leakage)	CMF	% of refrigerant flow rate
Improper outdoor airflow rate	CF	% of coil area blocked
Improper indoor airflow rate	EF	% of correct airflow rate
Liquid line restriction	LL	% of normal pressure drop
Refrigerant overcharge	OC	% overcharge from the correct charge
Refrigerant undercharge	UC	% undercharge from the correct charge

3.3.1 Compressor/reversing valve leakage

Compressor faults can arise from various reasons: gas leakage, improper lubrication, motor failure, etc. Brueker and Braun (1998) indicated that approximately 70 % of the classified faults are associated with a motor problem; motor performance degradation may be a result of overloading due to condenser fouling or a high outdoor temperature. Unstable electrical power - such as high/low voltage and voltage spike - can also cause motor problems.

The second major compressor fault is due to compressor valve leakage or other leakage which decreases the refrigerant mass flow rate. Wear and tear related to long-time operation of a reciprocating compressor may cause the refrigerant to leak through the suction or discharge valve. For scroll compressors, the refrigerant may leak in tangential directions through radial clearances between the neighboring pockets (flank leakage or tangential leakage) or in the radial direction through axial clearances between the rotating scroll and the body (tip leakage or radial leakage). Improper lubrication can degrade compressibility due to abrasion of contact surfaces like piston rings or cylinder walls. An internal bypass can arise from the intrusion of liquid refrigerant into the compressor. When the system starts up at low ambient temperatures or has the following faults: evaporator/condenser fouling, refrigerant overcharge, or excessively opened TXV, the compressor suction chamber may be flooded by liquid refrigerant. When the liquid refrigerant intrudes repeatedly into the compressor cylinder, mechanical parts like valves, rods, and piston will be damaged. Each of these faults degrades compression efficiency, whereas fatal compressor breakdown halts the entire system.

In this research we simulated an internal leak in the compressor by implementing a hot gas bypass shown in Figure 3.9. The bypass valve connected the compressor suction and discharge line close to the compressor and before the reversing valve was attached. For no-fault tests, the shut-off valve ensured no refrigerant flow through the bypass. During tests simulating a faulty compressor, fine tuning of the metering valve and a larger crude valve allowed establishment of a desired refrigerant flow rate through the bypass from the compressor discharge line to the suction line.

We expressed the severity of the artificial compressor leak as the reduction of refrigerant mass flow rate as compared to a no-fault condition. For each operating condition we performed a no-fault test first (with the shut-off valve closed) to obtain the reference mass flow rate. Then, we activated the bypass, and measured the new refrigerant mass flow rate, which was used for calculating the fault level. Refrigerant leakage through a four-way valve would affect the system performance similarly to the compressor leakage.

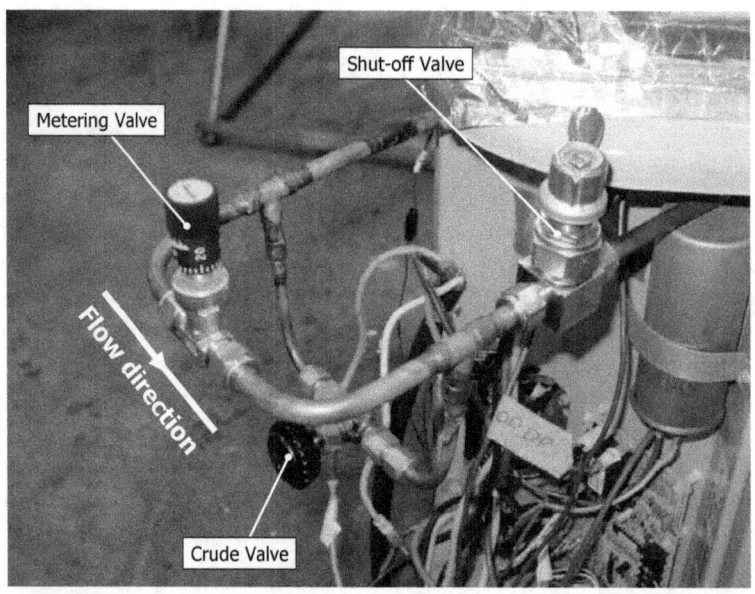

Figure 3.9. Hot gas bypass used for compressor/reversing valve leakage fault

3.3.2 Improper outdoor airflow rate

Outdoor sections are exposed to the outdoor environment and are easily contaminated by dirt or debris. Sometimes outdoor sections are surrounded by overgrown weeds or fallen leaves which restrict the airflow to the outdoor heat exchanger. The outdoor air flow may also decrease because of a defective fan motor, loose fan belt, or a poorly installed controller.

We simulated fouling of the outdoor heat exchanger by blocking the bottom part of its finned area with paper strips. The fault level was the percentage of the outdoor coil face area blocked by paper. Figure 3.10 shows the outdoor coil with a blockage or fault level of 35 %.

Figure 3.10. Outdoor unit with lower finned area blocked (35 % of the entire finned area blocked)

3.3.3 Improper indoor airflow rate

The airflow rate through the indoor section is affected by the size of ductwork, indoor fan sizing, and duct contamination. An improper duct design may burden the fan with an excessive load causing the fan to work below the nominal speed. Dust and debris collected on the heat exchanger may also result in a reduction in the airflow rate. Household articles like textile goods and carpets produce chemically reactive dust, and kitchens and baths also generate chemical vapor. If air filters are not maintained in good condition, these particles can flow into the ductwork and stick on the fan coil unit and duct walls reducing the airflow rate.

For this study the no-fault, reference air mass flow rate was set to 1699 $m^3\,h^{-1}$. For faulty tests, we reduced the speed of the nozzle chamber fan at the end of ductwork (see Figure 3.7). The fault level was the percent change in air mass flow rate with respect to the reference mass flow rate measured at no-fault conditions.

3.3.4 Liquid line restriction

Typically, a filter/dryer is installed in the liquid line to remove moisture and any solid particles from the circulating refrigerant. Moisture may enter the system if a service technician does not follow good refrigerant charging practices during servicing, while some particulates may exist in the system because of improper tube joinery technique. Accumulation of these substances will block the filter/dryer and unduly increase the refrigerant flow restriction.

To simulate an increased liquid line flow restriction, we installed a shut-off valve and metering valve in the liquid refrigerant line in a parallel configuration as shown in Figure 3.11 approximately 182 cm from the outdoor unit and within the outdoor chamber. By modulating the two valves, we controlled the liquid line restriction. The level of the liquid line restriction fault was the percent change in the liquid line pressure drop with respect to the pressure differential between the condenser exit and the evaporator inlet at the no-fault condition.

Figure 3.11. Artificial setup to implement a liquid line restriction fault using a shut-off valve and a metering valve

3.3.5 Refrigerant undercharge and overcharge

Residential systems are charge sensitive, i.e., their performance is influenced by the amount of refrigerant in the system. Refrigerant overcharge is a result of improper charging by a service technician. Refrigerant undercharge may result from improper charging or from a refrigerant leak. A rapid leak, caused by a component failure such as a fractured heat exchanger wall, is easy to detect because it degrades the system performance abruptly (a so called "hard fault"). A slow leak – e.g., due to a bad fitting in the refrigerant line where a small portion of refrigerant leaks for a long time – is typically difficult to detect, because the change in performance is slow and gradual (a so called "soft fault"). We simulated the overcharge and undercharge faults by adding or reducing the amount of refrigerant in the system. We determined the level of fault as the mass percentage of overcharged or undercharged refrigerant with respect to the optimized no-fault total refrigerant charge.

The improper charge was set by adding more or less refrigerant to the system with no change to the amount of POE (polyolester) oil. The charge level was established in reference to the nominal charge, assumed as 100%. A charge of 4.65 kg of R410A was taken as reference following the manufacturer's optimum charging criteria during the cooling mode.

3.4 Test Conditions

Table 3.3 presents operating conditions for the steady-state test series executed to map the performance of the system at normal (no-fault) operation and with imposed faults. For indoor conditions, the test program included three temperatures, 15.6 °C, 21.1 °C, and 23.9 °C. For outdoor conditions, we selected four temperatures: -8.3 °C, 1.67 °C, 8.3 °C, and 16.7 °C. Indoor relative humidity is not an influential parameter for performance of the condenser, and it was controlled roughly around 50 % within the range of 40 % to 60 %.

The test schedule for no-fault steady-state operation involved 13 indexed cases; three ARI 210/240 Standard rating tests, and 10 other tests. The fault tests were carried out for two operating conditions indicated in Table 3.3 by two asterisks (tests 3, 7 and 9). All no-fault tests were performed twice to check experimental repeatability, and steady-values are the average of these test results. In addition, a no-fault test preceded a series of fault tests at each of the operating conditions.

Table 3.3. Test conditions

Test index	Indoor		Outdoor	
	Dry-bulb temp. °C	Relative humidity %	Dry-bulb temp. °C	Relative humidity %
1*	21.11	40 to 60	-8.33	67
2*	21.11	40 to 60	1.67	82
3*,**	21.11	40 to 60	8.33	73
4	15.56	40 to 60	-8.33	Dry
5	15.56	40 to 60	1.67	Dry
6	15.56	40 to 60	8.33	Dry
7**	21.11	40 to 60	-8.33	Dry
8	21.11	40 to 60	1.67	Dry
9**	21.11	40 to 60	8.33	Dry
10	21.11	40 to 60	16.67	Dry
11	23.89	40 to 60	-8.33	Dry
12	23.89	40 to 60	16.67	Dry
13	21.11	40 to 60	16 to 3	50 to 78

* ARI Standard 210/240 (2006)
** Combination of test conditions selected for fault tests

CHAPTER 4. Fault-Free Tests and System Characteristics

The unit was tested in the heating mode after system refrigerant charge and proper operating characteristics were confirmed in the cooling mode. Standard heating mode rating condition tests, fault-free steady-state tests, fault-free transient tests, frosting tests, and fault-free repeatability tests were performed.

4.1 Standard Rating Condition Test Results

Table 4.1 compares the published ratings for this unit to our test results. The heating mode performance of the unit matched closely the published ratings. The greatest difference occurred during the low temperature heating test, but the test results are very close to the uncertainty limits for capacity at these low capacity heating conditions. System to system variability could also explain the lower tested capacity at these low temperature heating conditions.

Table 4.1. Performance data for AHRI rating tests

	High Temp. Heating Capacity (W) [1]	Low Temp. Heating Capacity (W) [2]	High Temp. Heating Power (W)	*Low Temp. Heating Power (W)	*High Temp. Heating COP	*Low Temp. Heating COP
Measurements	8441	5275	2435 (total) 1910 (comp)	2325	3.46	2.26
Test Results	(Air) 8223 -2.6 % (Ref) 8606 2.0 % Energy Balance [3] 4.7 %	(Air) 4908 -7.0 % (Ref) 5022 -4.8 % Energy Balance 2.3 %	2361 (total) 1798 (comp)	2181 (total) 1594 (comp)	3.48 (Air) 3.65 (Ref)	2.25 (Air) 2.30 (Ref)

(Air) air side measurement, (Ref) refrigerant side measurement (measured fan heat subtracted from capacity), (comp) compressor
1) High temperature heating condition: outdoor 8.33 °C DB/6.11 °C WB, indoor 21.11 °C DB
2) Low temperature heating condition: outdoor -8.33 °C DB/-9.44 °C WB, indoor 21.11 °C DB
3) Energy balance = 100 x (Air side capacity – Ref. side capacity)/(Air side capacity)

4.2 Fault-Free Steady-State Test Results

The effects of frosting were avoided during the fault-free testing by maintaining a low outdoor dew point temperature. These tests ensured consistent results from test to test and removed the variability in system performance associated with frost formation. All fault-free steady-state tests presented in this section were performed at frost-free conditions around the outdoor unit.

4.2.1 System characteristics

Figure 4.2.1.1 shows the air side heating capacity (including fan heat) and COP of the system as a function of the outdoor ambient conditions. Refrigerant side capacity of the outdoor heat exchanger is also shown along with compressor power.

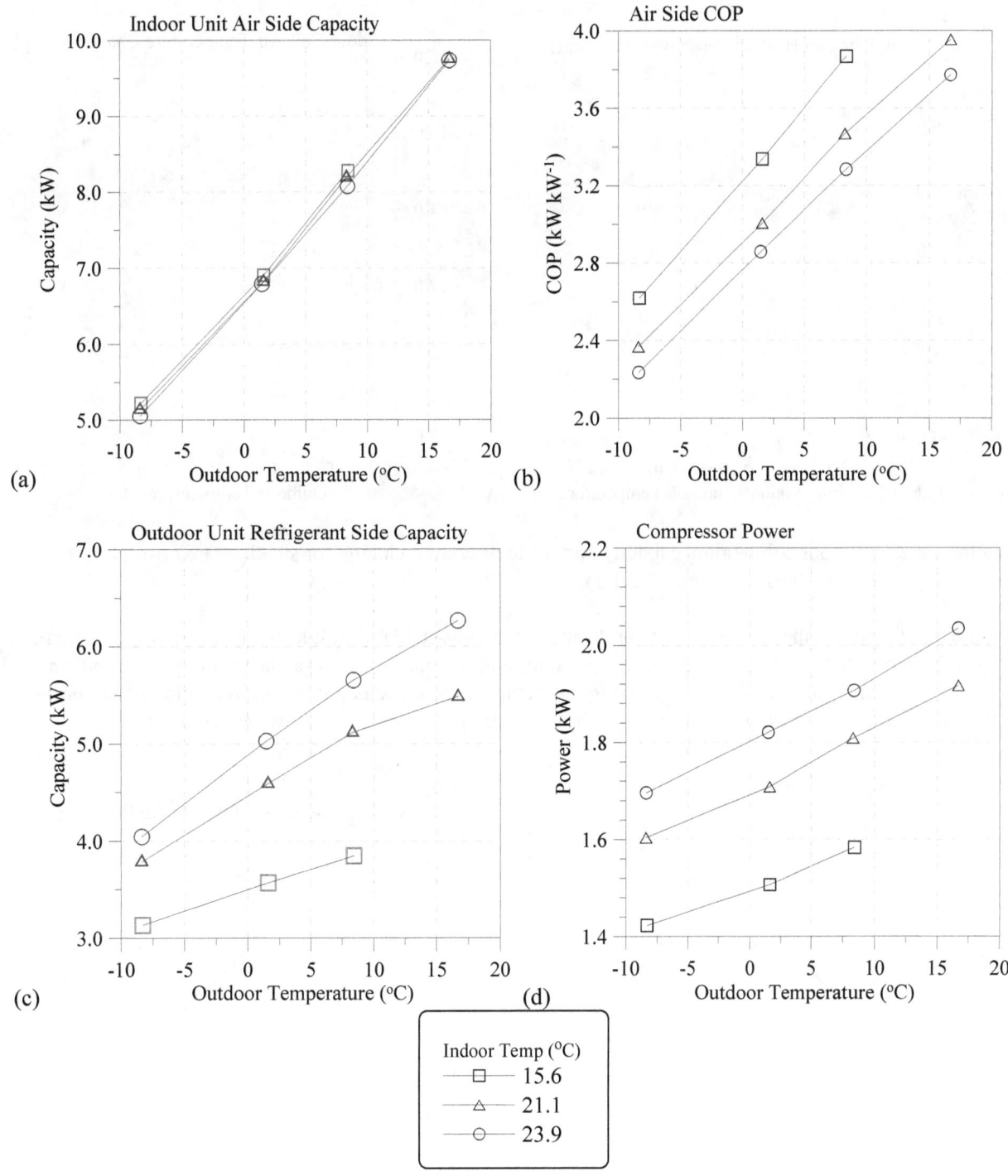

Figure 4.2.1.1. Capacity, COP, and compressor power for steady-state no-fault tests

4.2.2 Indoor unit characteristics

As shown in Figure 4.2.2.1(a), the indoor unit capacity is not a strong function of indoor temperature. Figure 4.2.2.1(a) shows indoor coil alone heating capacity as a function of the refrigerant vapor inlet saturation temperature at the three indoor ambient temperature conditions. Coil heating capacity is a linear function of the refrigerant saturation temperature at all conditions shown.

19

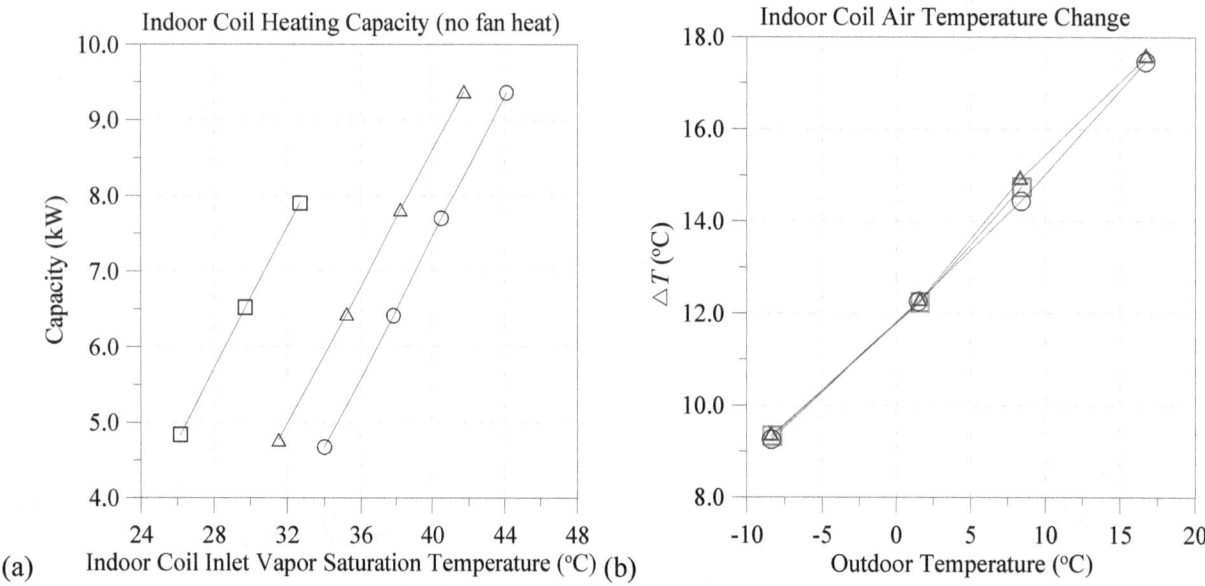

(a)   (b)

Figure 4.2.2.1. Indoor coil heating capacity and air temperature change for steady-state no-fault tests (for symbols see Figure 4.2.1.1)

Figure 4.2.2.2 shows the indoor fan power during the steady-state no-fault tests. Figure (b) shows that fan power is very nearly a linear function of the temperature measured by a thermocouple placed on the exterior of the fan motor casing. This casing temperature provides a convenient indication of the no-fault amperage draw of the indoor fan motor at the given static pressure and air flow rate conditions.

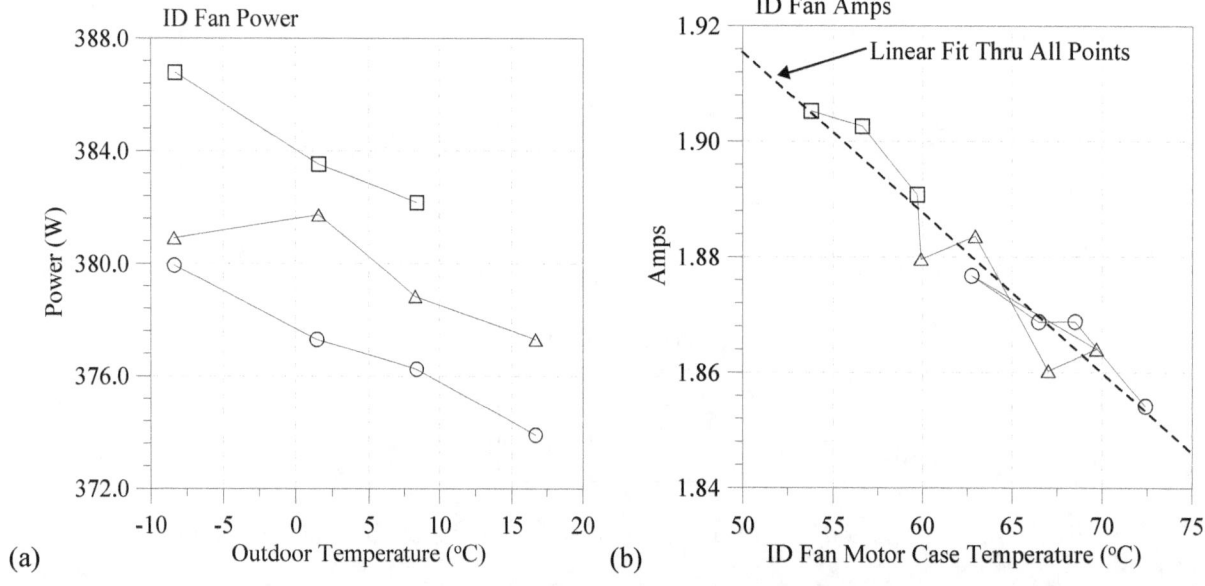

(a)   (b)

Figure 4.2.2.2. Indoor unit fan power and amps during heating mode steady-state no-fault tests (see Figure 4.2.1.1 for symbols)

Figure 4.2.2.3 shows the indoor unit standard air flow rate and the air flow rate per unit heating capacity. The indoor fan operates at constant speed thus more mass flow is produced at lower indoor temperatures.

At a given indoor temperature the indoor air flow decreases with increasing outdoor temperature due to higher temperature changes (more capacity and less dense air) at higher outdoor temperatures. Air flow rate per unit of heating capacity is not a function of indoor temperature.

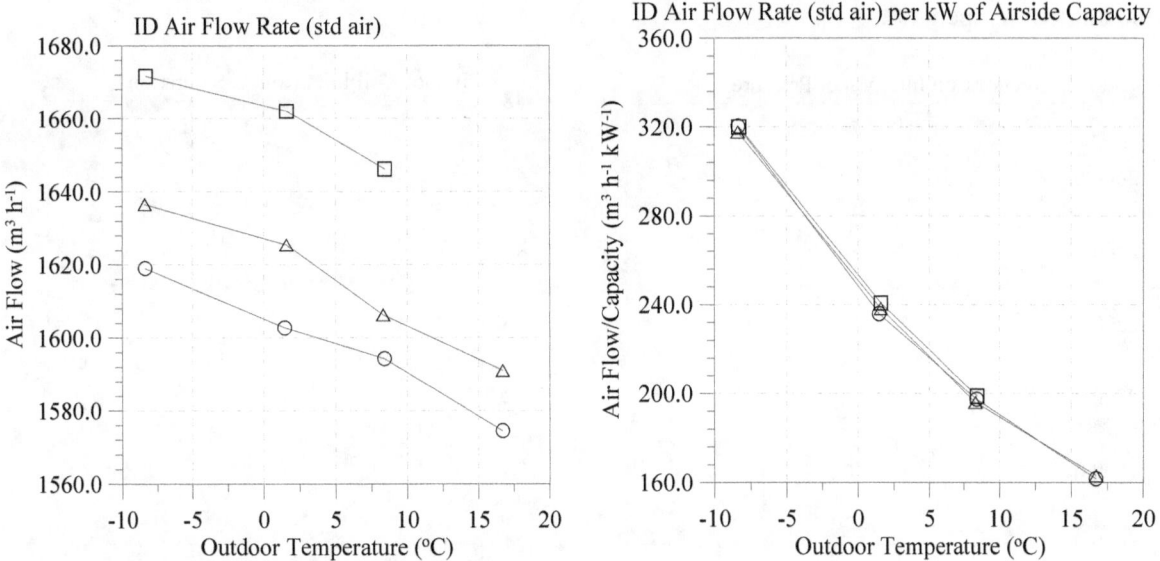

Figure 4.2.2.3. Indoor unit air flow rate and capacity-normalized air flow for steady-state no-fault tests (see Figure 4.2.1.1 for symbols)

Figure 4.2.2.4 shows the inlet and exit refrigerant conditions and refrigerant pressure drop through the condenser. The condenser refrigerant-side pressure drop is not a function of the indoor temperature and varies almost linearly with outdoor temperature. The condenser saturation temperature varies linearly with the outdoor temperature.

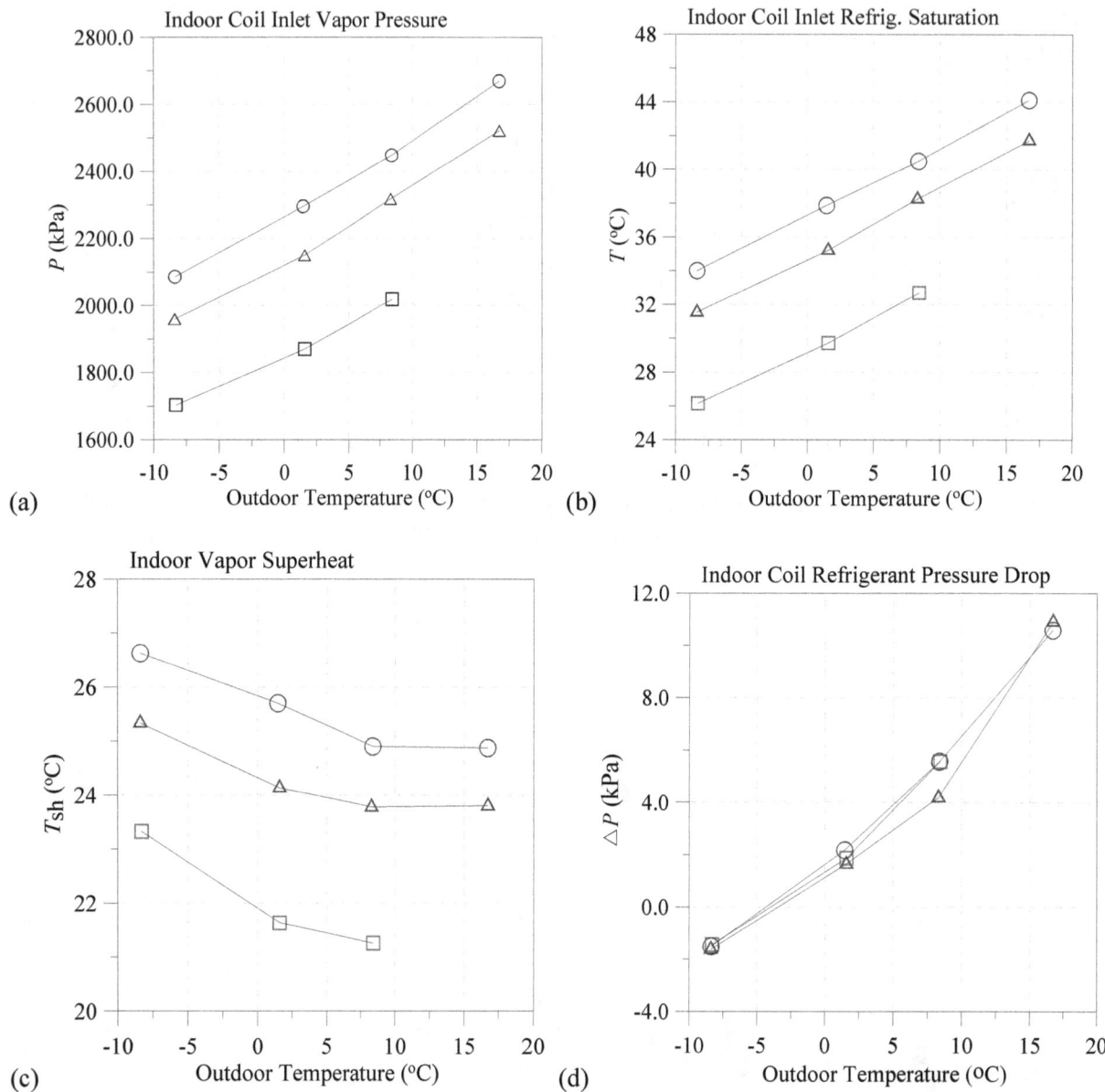

Figure 4.2.2.4. Condenser or indoor coil refrigerant measurements for steady-state no-fault tests (see Figure 4.2.1.1 for symbols)

Figure 4.2.2.5 shows the location of various thermocouples attached to the indoor heat exchanger. Figure 4.2.2.6 compares temperature measured at locations #3, #5, #12, and #15 with the refrigerant saturation temperature corresponding to the pressure at the condenser inlet. As figures (a) and (b) indicate, tube bend temperatures after the first pass are indicative of the saturation temperature at the condenser inlet pressure.

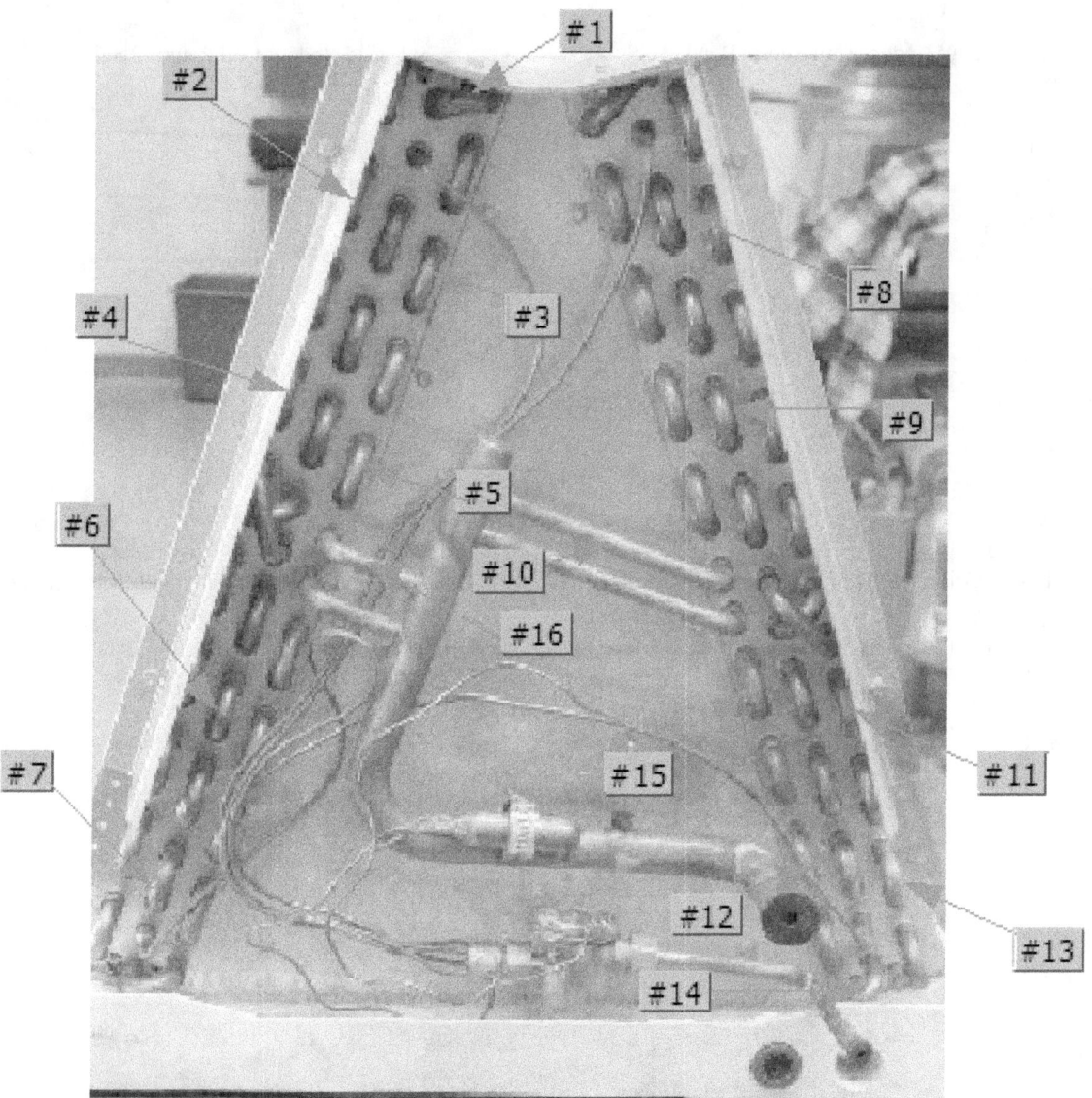

Figure 4.2.2.5. Thermocouple locations on the indoor heat exchanger

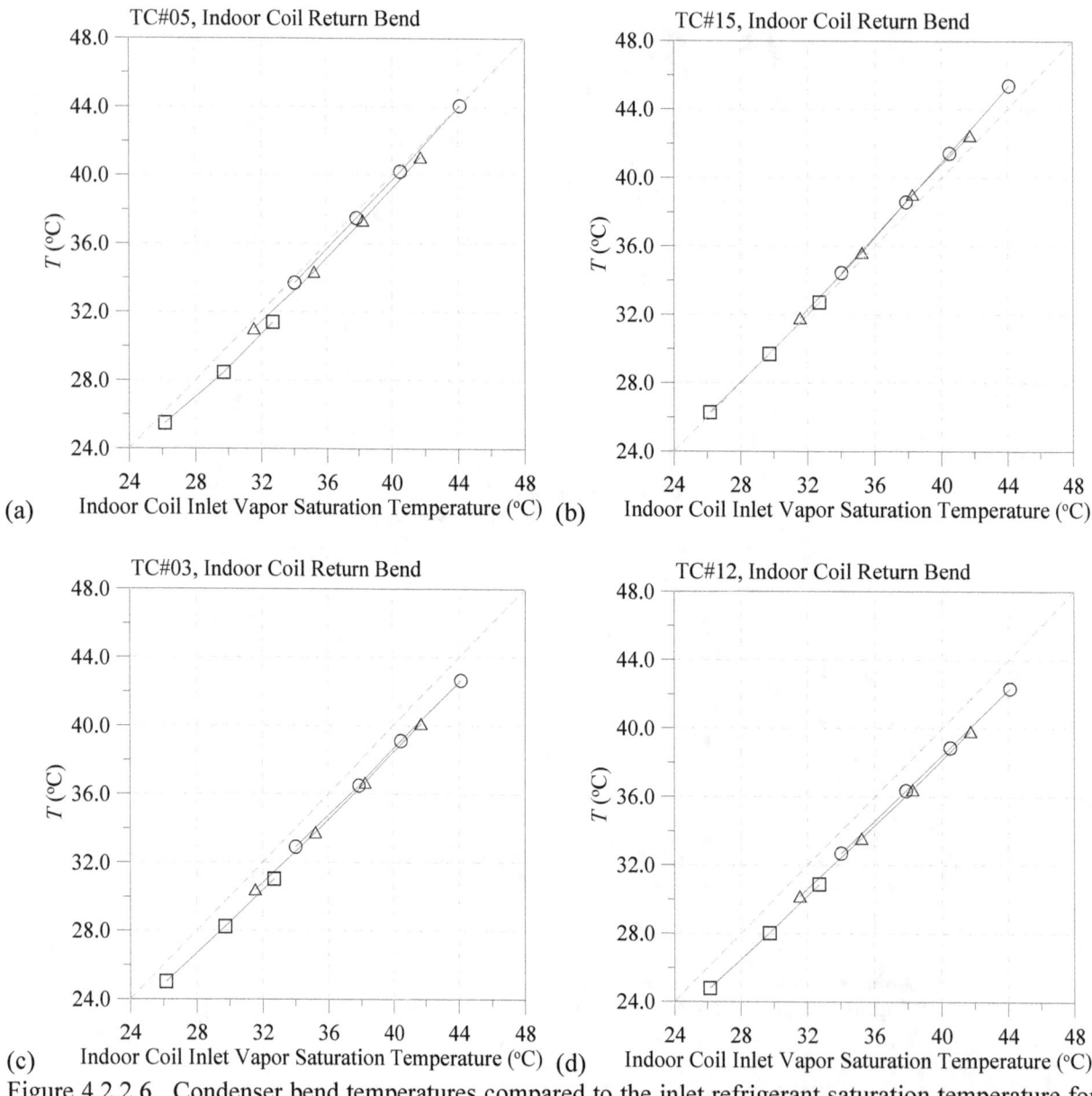

Figure 4.2.2.6. Condenser bend temperatures compared to the inlet refrigerant saturation temperature for steady-state no-fault tests (see Figure 4.2.1.1 for symbols)

4.2.3 Outdoor unit characteristics

Figure 4.2.3.1 through 4.2.3.5 shows different compressor characteristics during No-Fault Steady-State, NFSS, tests at three indoor temperatures.

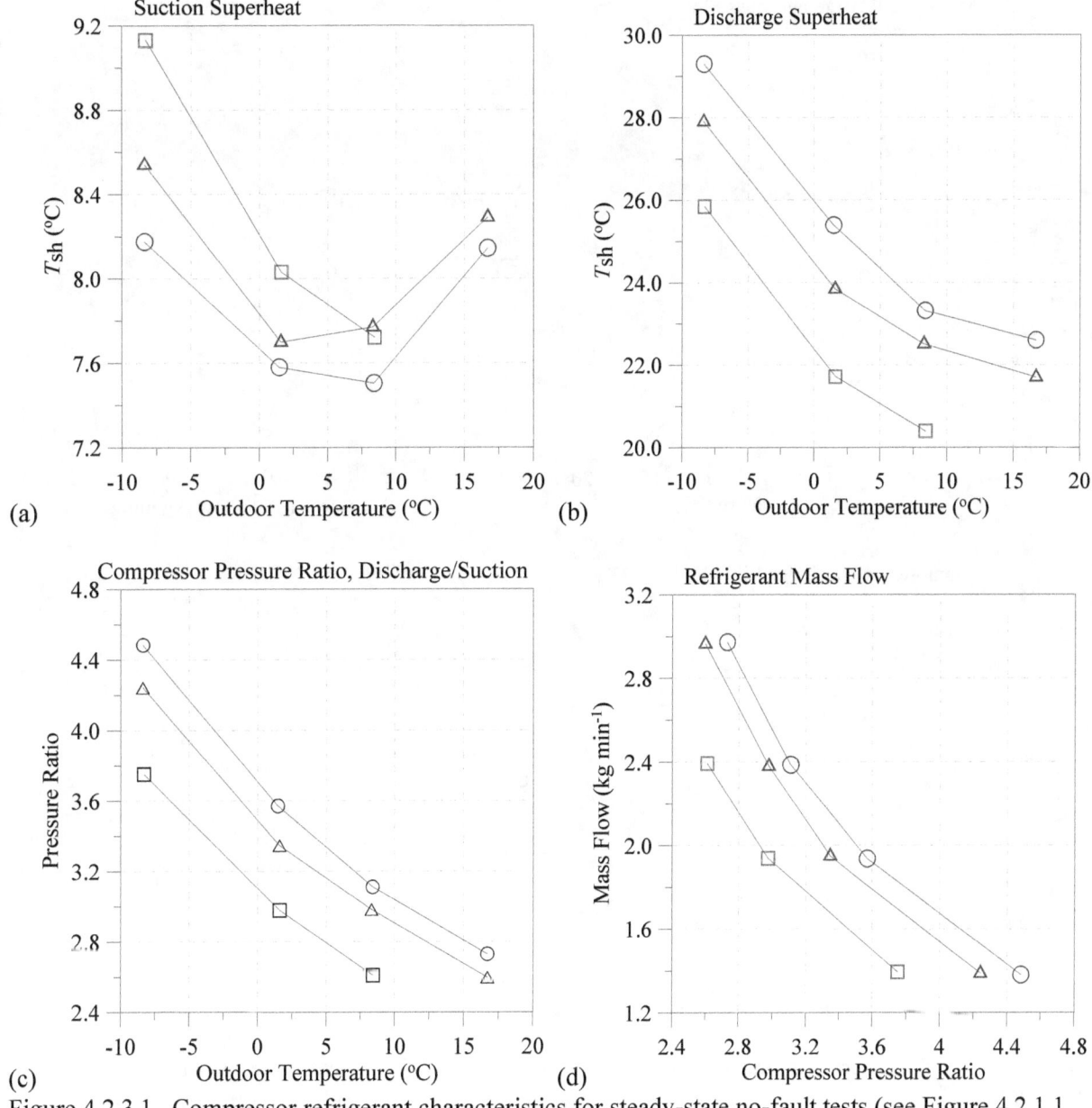

Figure 4.2.3.1. Compressor refrigerant characteristics for steady-state no-fault tests (see Figure 4.2.1.1 for symbols)

Figure 4.2.3.2 shows that the compressor map mass flow rate was within -5 % of the measured mass flow. This is very good agreement and not typical of all compressors. Discharge and suction temperatures as a function of outdoor temperature are also shown.

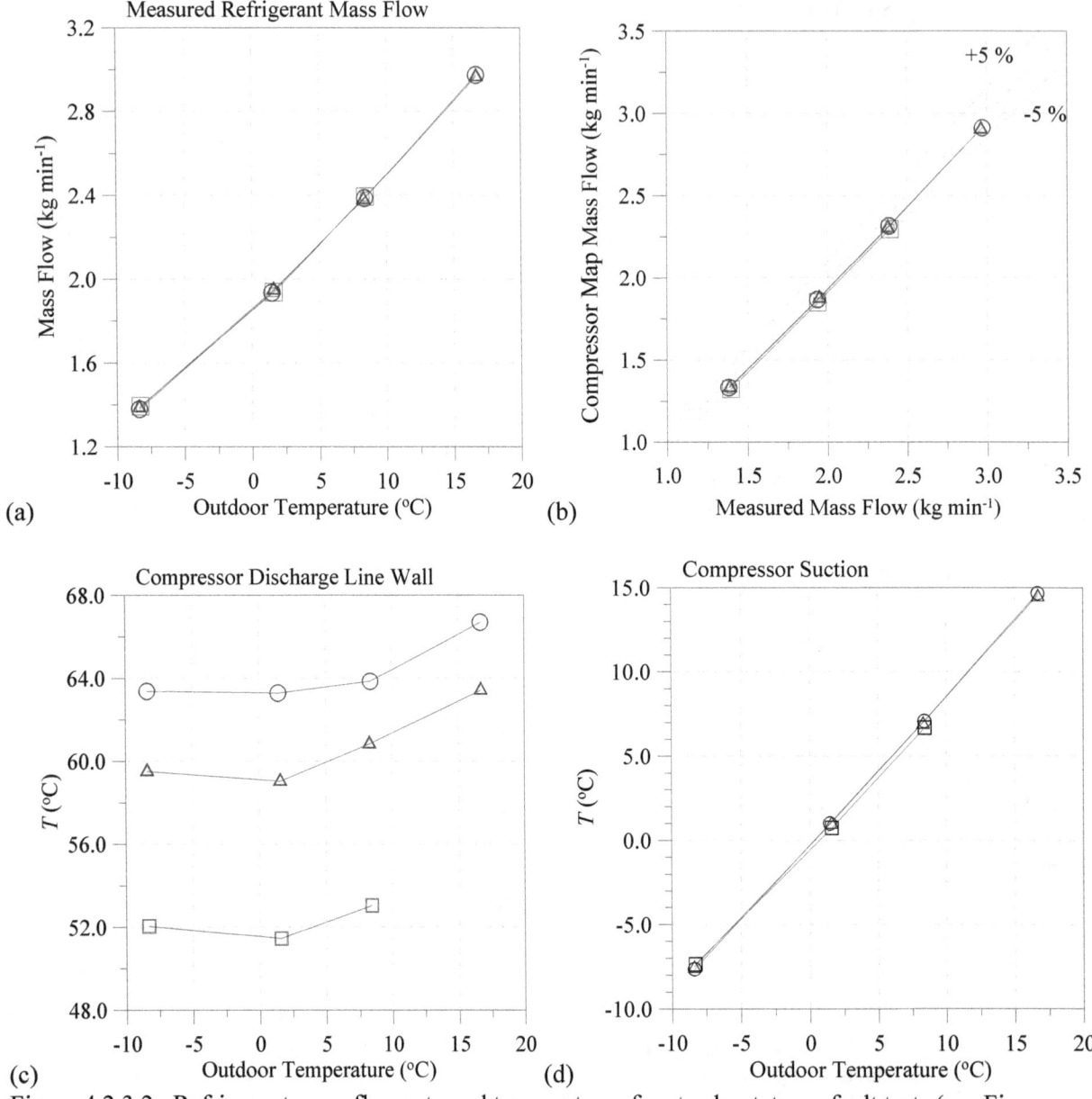

Figure 4.2.3.2. Refrigerant mass flow rate and temperatures for steady-state no-fault tests (see Figure 4.2.1.1 for symbols)

Compressor power and casing temperature characteristics are shown in Figure 4.2.3.3. Power and current demand are very linear with outdoor temperature for a given indoor temperature. The compressor current appears to correlate with compressor shell top temperature, especially at temperatures above 0 °C; above 0 °C the amps closely approach a linear function of shell top temperature. Figure 4.2.3.4 shows, as expected, that compressor amps correlate to pressure ratio and suction/discharge temperature difference across the compressor.

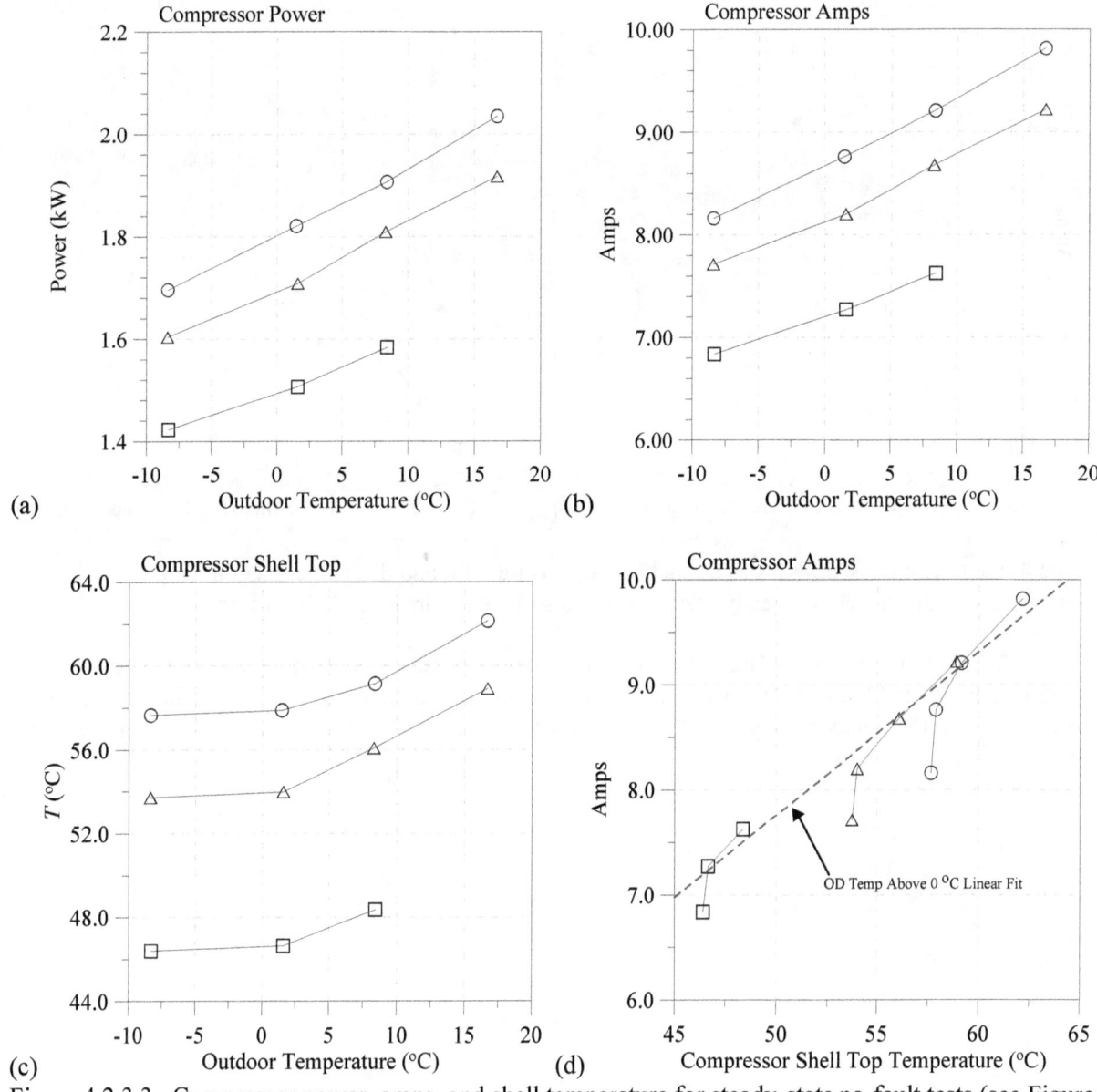

Figure 4.2.3.3. Compressor power, amps, and shell temperature for steady-state no-fault tests (see Figure 4.2.1.1 for symbols)

27

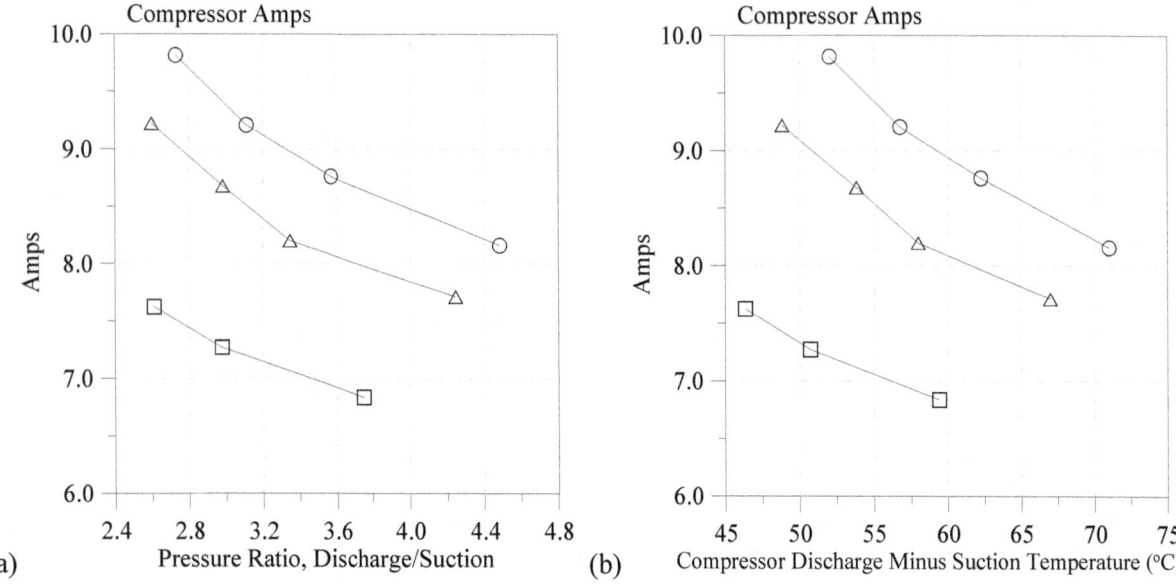

(a)   Pressure Ratio, Discharge/Suction   (b)   Compressor Discharge Minus Suction Temperature (°C)

Figure 4.2.3.4. Compressor amps as a function of pressure ratio and discharge-suction temperature difference for steady-state no-fault tests (see Figure 4.2.1.1 for symbols)

Figure 4.2.3.5 shows refrigerant side capacity and general characteristics of the outdoor heat exchanger. Figure (d) shows that the evaporator exit refrigerant saturation temperature was below the outdoor dew-point for the 21.1 °C indoor temperature tests; condensation could form upon the coil for some of these tests.

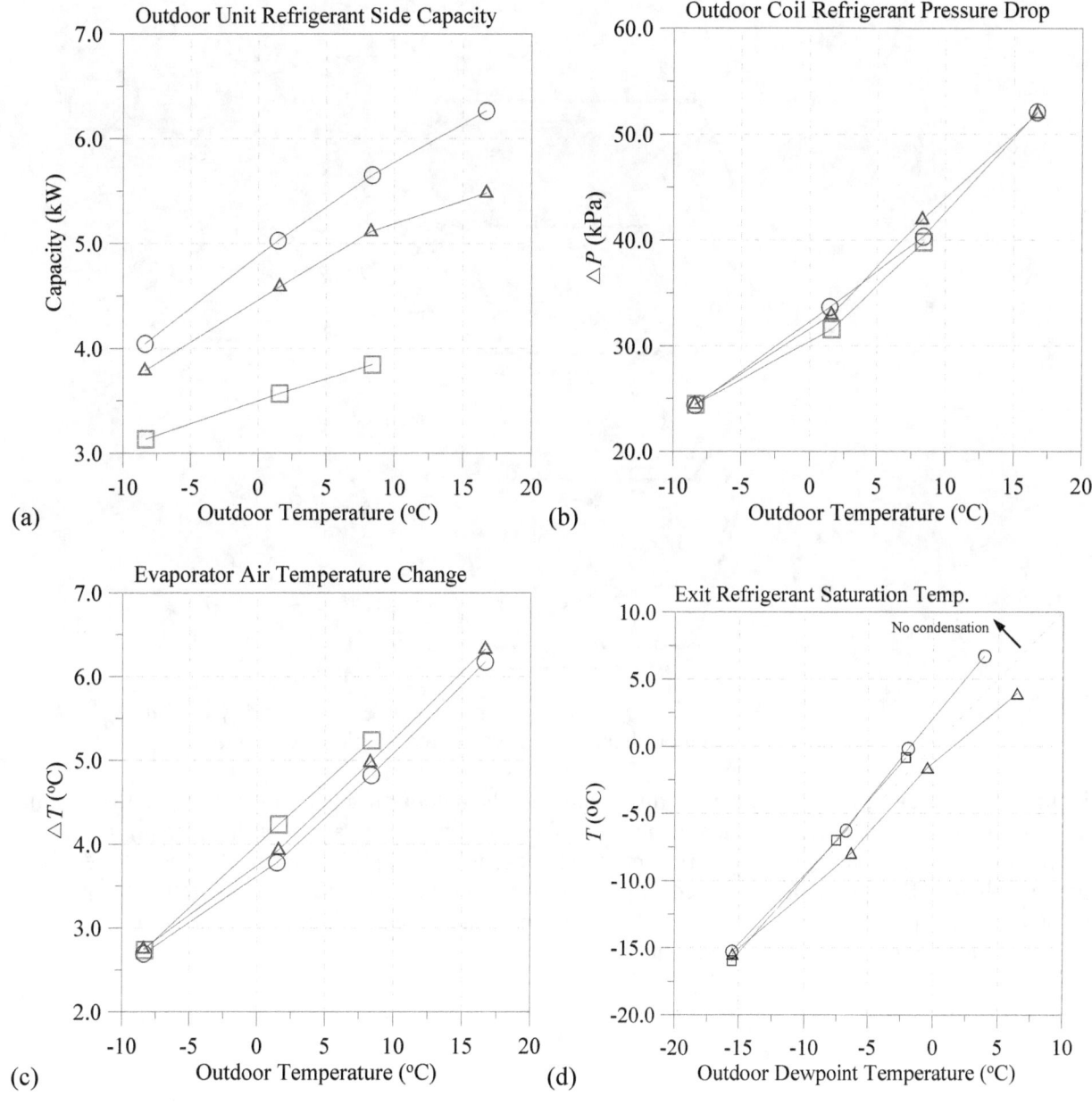

Figure 4.2.3.5. Evaporator refrigerant side capacity, pressure drop, and evaporator temperatures for steady-state no-fault tests (see Figure 4.2.1.1 for symbols)

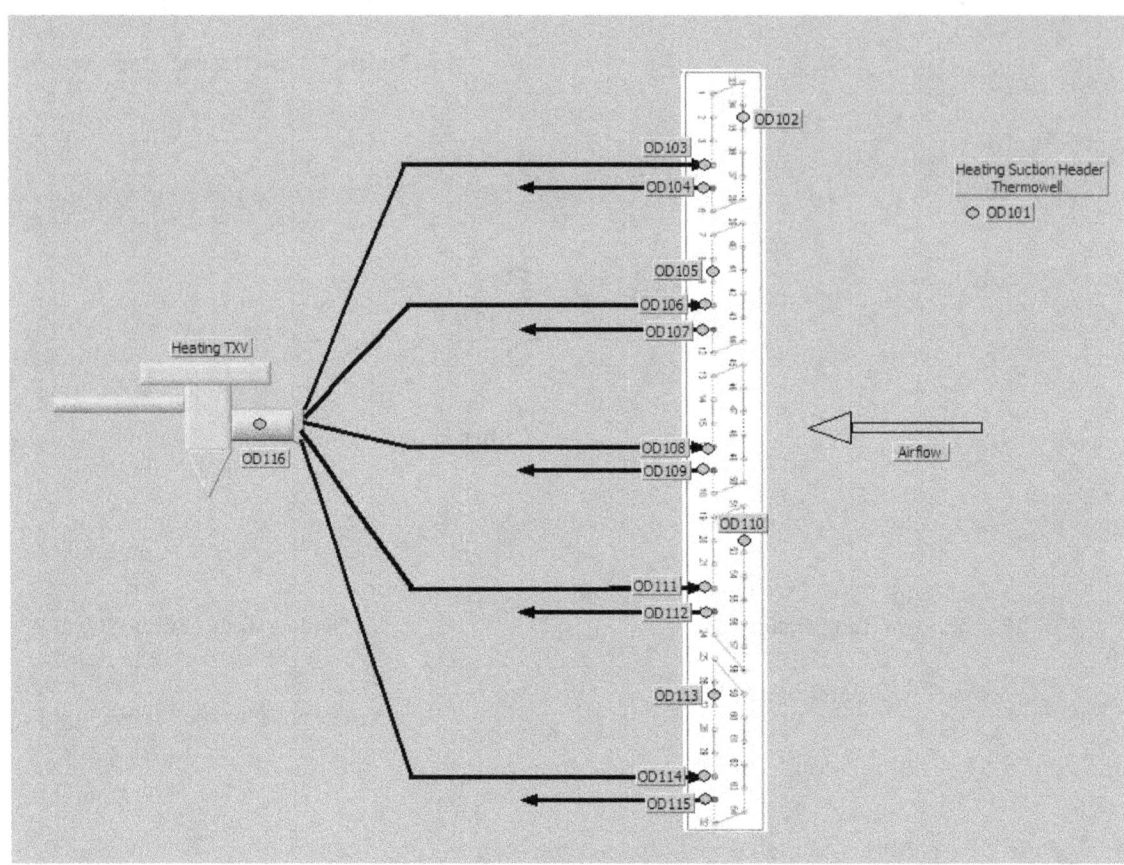

Figure 4.2.3.6. Outdoor heat exchanger thermocouple placements

Figure 4.2.3.6 shows the placement of thermocouples for the outdoor heat exchanger (evaporator). Figure 4.2.3.7 compares evaporator bend temperatures to the exit refrigerant saturation temperature calculated from a measured pressure. The measurements taken close to the evaporator inlets all appear to represent the two-phase refrigerant temperature equally and differ from the evaporator exit saturation temperature by an average offset of +2.1±0.3 °C. Any of these temperatures could be used to represent the evaporator exit saturation temperature using a constant offset.

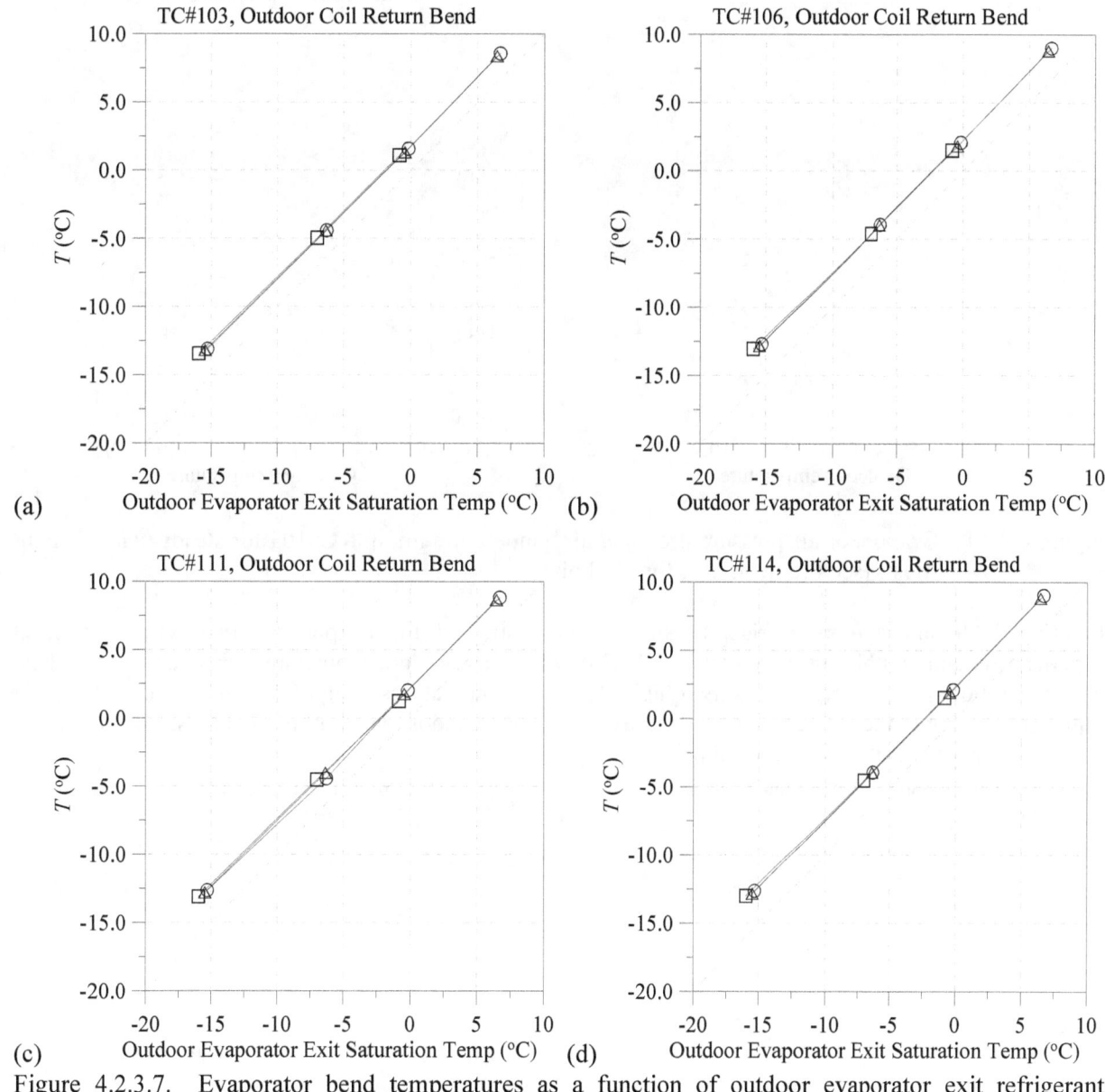

Figure 4.2.3.7.   Evaporator bend temperatures as a function of outdoor evaporator exit refrigerant saturation temperature (see Figure 4.2.1.1 for symbols)

Figure 4.2.3.8 shows air side characteristics of the outdoor heat exchanger. Air pressure drop across the outdoor heat exchanger was relatively constant for a given outdoor temperature. Any frost formation would be indicated by a higher air pressure drop, thus this indicates absence of frost on the heat exchanger. The temperature difference indicated by figure (b) is also a good indicator of frost formation, and as the figure shows, this temperature difference varies linearly with outdoor and indoor temperature.

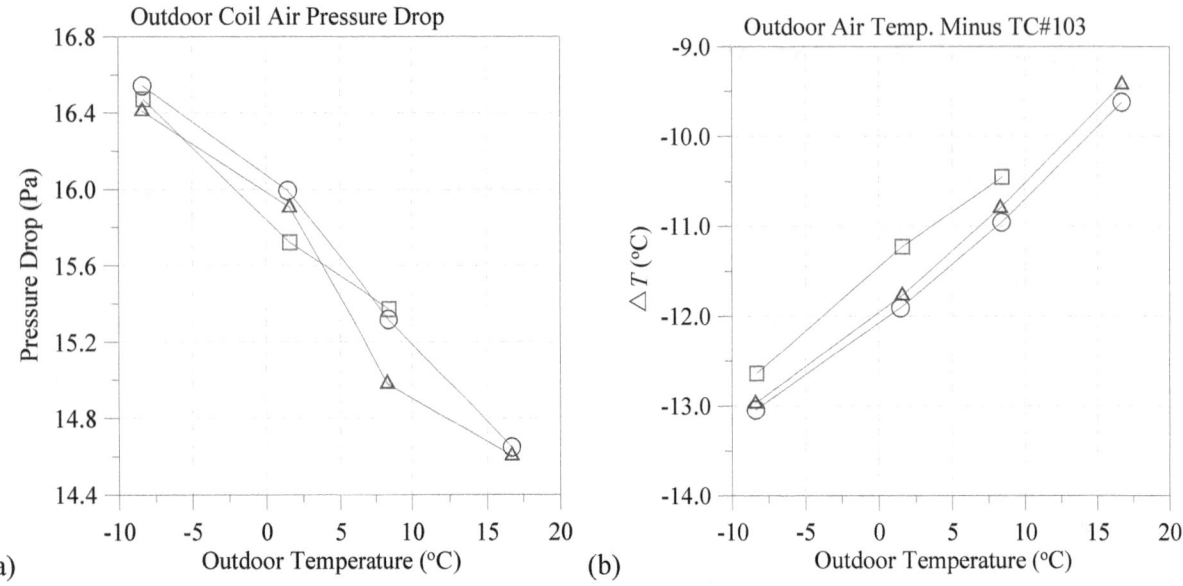

Figure 4.2.3.8. Evaporator air pressure drop and air temperature minus TC#103 for steady-state no-fault tests (see Figure 4.2.1.1 for symbols)

Figure 4.2.3.9 shows more refrigerant side characteristics of the evaporator. The exit refrigerant superheat remains within the range from 5.8 °C to 7.8 °C as outdoor temperature increases. Superheat does not follow a linear trend at the two highest indoor temperatures. Figure (b) shows that refrigerant liquid subcooling at the service valve decreases linearly as outdoor temperature increases; higher heating capacity correlates with lower subcooling.

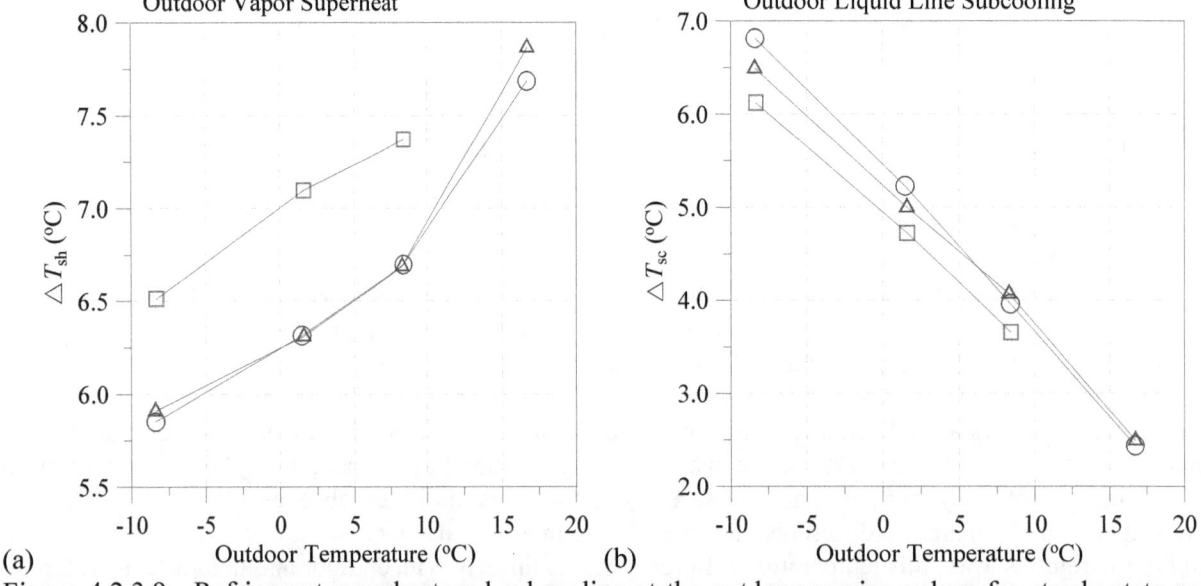

Figure 4.2.3.9. Refrigerant superheat and subcooling at the outdoor service valves for steady-state no-fault tests (see Figure 4.2.1.1 for symbols)

Figure 4.2.3.10 shows the power characteristics of the outdoor fan motor. Figures (b) and (c) show that the fan motor amp requirements correlate well with the outdoor ambient temperature and the temperature measured by a thermocouple placed on the exterior of the fan motor casing. Using the fan motor case and outdoor ambient temperature difference produces a slightly better correlation of outdoor fan motor amps than using the outdoor ambient temperature alone (28 % lower residual sum of squares).

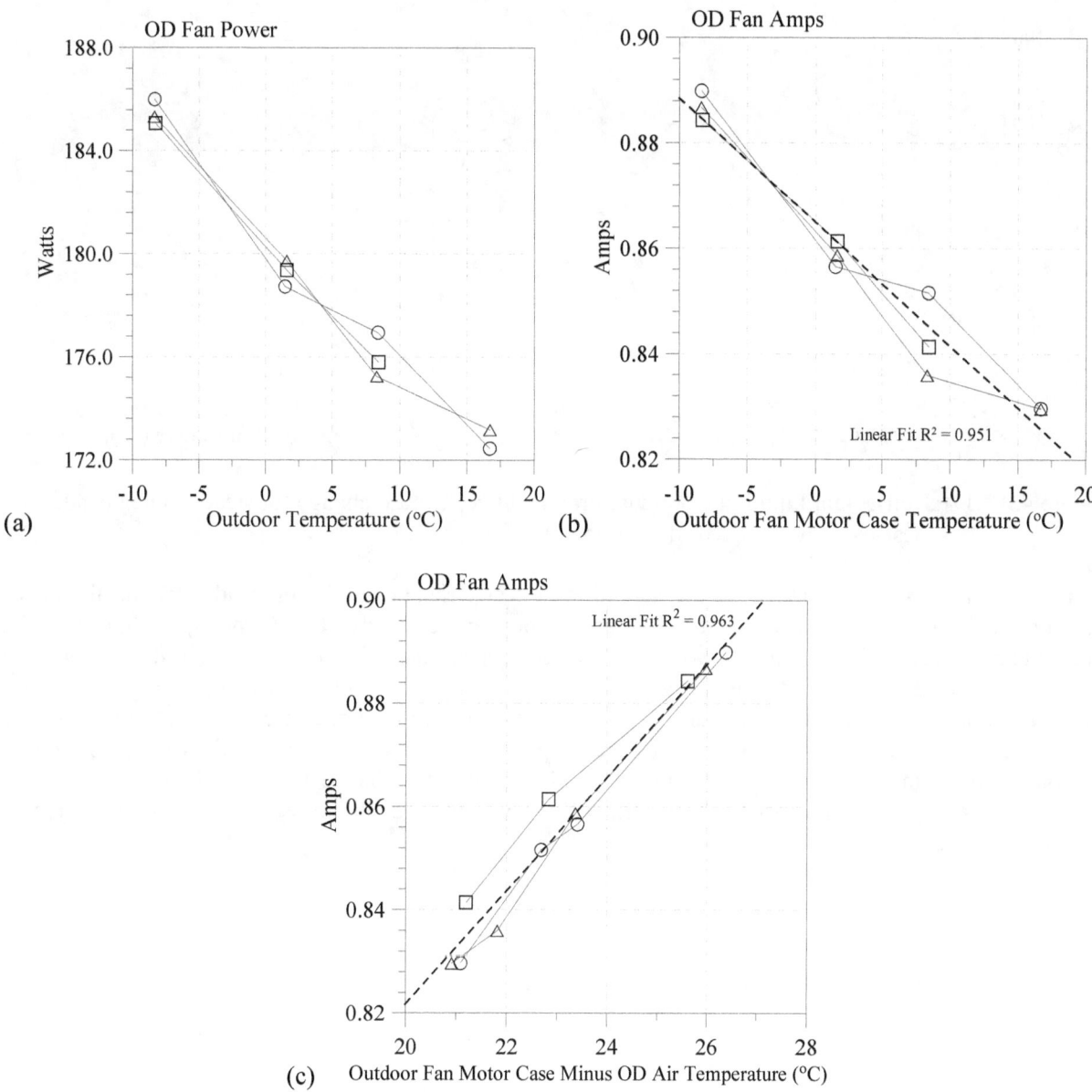

Figure 4.2.3.10. Outdoor fan power characteristics for steady-state no-fault tests (see Figure 4.2.1.1 for symbols)

Figure 4.2.3.11 shows the refrigerant characteristics for the liquid line used with this system. Figure (a) shows refrigerant pressure drop through the liquid line from the outdoor unit service valve to the indoor unit's liquid connection. This pressure drop is a smooth, although non-linear, function of the outdoor temperature and a weak function of indoor temperature. Figure (b) shows that the corresponding temperature change in the liquid line was greatest at the lower outdoor temperature, higher subcooling, lower capacity conditions. This temperature change is affected by refrigerant mass flow rate and thermal contact with the surrounding environment.

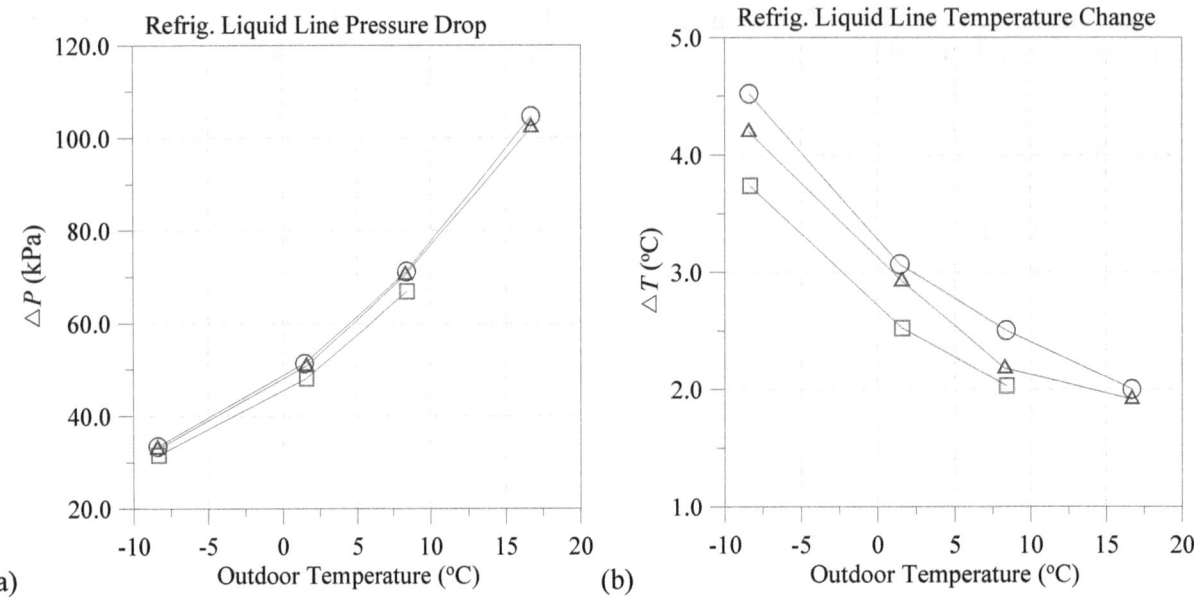

(a)　　　　　　　　　　　　　　　　　　　　(b)

Figure 4.2.3.11. Refrigerant liquid line pressure drop and temperature change for steady-state no-fault
tests(see Figure 4.2.1.1 for symbols)

Figure 4.2.3.12 shows reversing valve pressure drops and superheat changes for steady-state no-fault tests. Pressure drops have the greatest variability at an outdoor temperature of 8.3 °C, being the highest at an indoor temperature of 21.1 °C and lowest at an indoor temperature of 15.6 °C. For all these tests at an outdoor temperature of 8.3 °C, the refrigerant mass flow rate was fairly constant, 2.4±0.05 kg min^{-1}. Tests at indoor temperatures other than 21.1 °C have a much more linear pressure drop with outdoor temperature. The reason for the higher pressure drop at 21.1 °C indoor/8.3 °C outdoor does not appear obvious at this time, but divergence from a smooth temperature change (seen in Figures (c)) is also present. This may suggest a bad valve seating at this condition, but the mass flow data did not suggest this was the case.

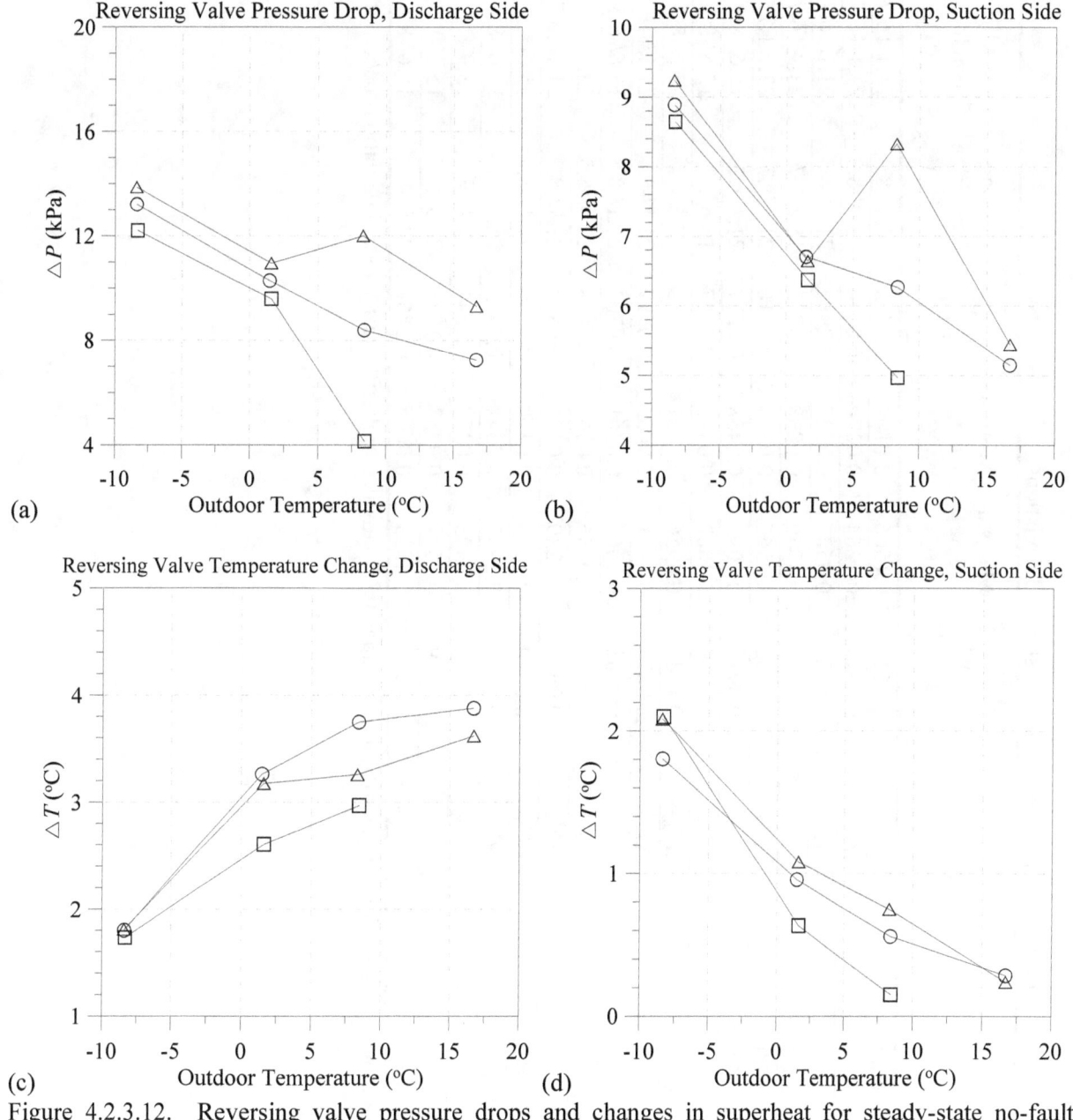

Figure 4.2.3.12. Reversing valve pressure drops and changes in superheat for steady-state no-fault tests(see Figure 4.2.1.1 for symbols)

4.2.4 Steady-state feature standard deviations

The standard deviations of system features are necessary for development of the steady-state detector, as presented for the cooling mode by Kim et al. 2006. Table 4.2.4.1 lists selected features along with their measured standard deviations averaged over all of the steady-state no-fault tests.

Table 4.2.4.1. Selected feature standard deviations

Feature name	Feature symbol	Average standard deviation (°C)	Max std. dev. (°C)	Min std. dev. (°C)	Range (°C)	Max condition* (°C)	Min condition* (°C)
evaporator exit saturation temperature	T_{ER}	0.0645	0.1566	0.0101	0.1465	21.1/-8.3	21.1/16.7
evaporator bend thermocouple, TC#103	T_{E103}	0.0644	0.1264	0.0370	0.0894	23.9/-8.3	21.1/16.7
evaporator exit superheat	ΔT_{shE}	0.1127	0.3507	0.0525	0.2982	21.1/-8.3	21.1/16.7
compressor discharge wall temperature	T_D	0.1779	0.4740	0.0868	0.3872	15.6/1.7	23.9/16.7
condenser inlet saturation temperature	T_{CR}	0.0750	0.2567	0.0358	0.2210	21.1/1.7	23.9/1.7
condenser bend thermocouple, TC#15	T_{C15}	0.0913	0.3095	0.0458	0.2637	21.1/1.7	15.6/-8.3
condenser inlet superheat	ΔT_{shC}	0.1469	0.4882	0.0415	0.4467	15.6/1.7	15.6/-8.3
vapor superheat at outdoor service valve	ΔT_{shV}	0.1946	0.4956	0.0521	0.4435	15.6/1.7	15.6/-8.3
liquid line subcooling at outdoor service valve	ΔT_{scV}	0.0874	0.1451	0.0653	0.0799	21.1/-8.3	21.1/8.3
condenser air temperature rise	ΔT_{CA}	0.0588	0.1064	0.0361	0.0703	21.1/-8.3	15.6/-8.3
evaporator air temperature drop	ΔT_{EA}	0.0562	0.0856	0.0367	0.0488	23.9/-8.3	23.9/16.7
liquid line temperature drop	ΔT_{LL}	0.0687	0.1239	0.0436	0.0803	21.1/-8.3	21.1/16.7
outdoor temperature minus TC#103	ΔT_{103}	0.0680	0.1509	0.0378	0.1130	21.1/-8.3	15.6/8.3
ID fan motor case temperature	T_{IDF}	0.1392	0.2957	0.0544	0.2413	21.1/1.7	15.6/-8.3
OD fan motor case temperature	T_{ODF}	0.0750	0.2303	0.0259	0.2044	21.1/-8.3	23.9/1.7
reversing valve temperature change, discharge side	ΔT_{RVD}	0.1409	0.2577	0.0753	0.1823	23.9/16.7	15.6/-8.3
reversing valve temperature change, suction side	ΔT_{RVS}	0.1576	0.4533	0.0665	0.3868	23.9/-8.3	15.6/-8.3

*Max and Min condition refers to the indoor/outdoor dry-bulb temperatures at which highest/lowest standard deviations occurred.

4.3 Fault-Free Start-Up Transient Test Results

The essential part of the FDD methodology under development in this study is a steady-state detector. As the first step in developing this steady-state detector, start-up testing was performed to determine the time associated with the system reaching steady-state after being off for more than 30 minutes. The methodology presented here follows that presented by Kim et al. (2008b). The figures presented below are intended to show steady-state operation after the moving window standard deviation of the feature falls below and stays below three times its NFSS value. The tests were performed at a single indoor dry-bulb temperature of 21.1 °C and outdoor dry-bulb temperatures of -8.3 °C, 1.7 °C, 8.3 °C, and 16.7 °C for dry evaporator conditions. Of the features listed above in Table 4.2.4.1, standard deviations during steady-state testing were the smallest for an indoor/outdoor dry-bulb temperature of 21.1/16.7 °C and the largest at 21.1/-8.3 °C. The term "scan(s)" refers to a single measurement or sampling by the data acquisition system.

4.3.1 Start-up variation of features at an indoor/outdoor dry-bulb temperature of 21.1 °C /16.7 °C

Table 4.3.1.1 presents the time required for the moving window standard deviations of the selected features to remain below a three standard deviation line. The same information is presented graphically in Figures 4.3.1.1 through 4.3.1.7. The solid horizontal line in each figure indicates three standard deviations measured during the fault-free steady-state tests at the given ambient conditions. Underlined entries in Table 4.3.1.1 indicate that the feature's moving window standard deviation, for that moving window scan size, fell below the three standard deviation line and rose above it again before remaining below; the feature oscillated around the three standard deviation line.

Table 4.3.1.1. Time required for features to remain below three standard deviations for various moving window sample sizes at an indoor/outdoor dry-bulb temperature of 21.1 °C /16.7 °C

Feature Name	Feature Symbol	Three Scans	Four Scans	Five Scans
		minutes		
total air side capacity	Q_{CA}	6.00	7.00	7.50
compressor amps	-	3.50	4.20	4.80
refrigerant mass flow rate	m_R	4.80	5.40	6.00
evaporator exit saturation temperature	T_{ER}	-	-	-
evaporator bend thermocouple, TC#103	T_{E103}	4.75	8.93	8.93
evaporator exit superheat	ΔT_{shE}	7.15	7.15	7.75*
compressor discharge wall temperature	T_D	7.75	8.33	8.33
condenser inlet saturation temperature	T_{CR}	4.77	5.37	5.95
condenser bend thermocouple, TC#15	T_{C15}	5.37	5.95	6.55
condenser inlet superheat	ΔT_{shC}	5.37	5.95	6.55
vapor superheat at outdoor service valve	ΔT_{shV}	5.37	5.95	6.55
liquid line subcooling at outdoor service valve	ΔT_{scV}	4.17	3.57	4.17
condenser air temperature rise	ΔT_{CA}	7.15	8.33	8.93
evaporator air temperature drop	ΔT_{EA}	2.98	3.57	4.17
liquid line temperature drop	ΔT_{LL}	3.57	4.77	4.77
outdoor temperature minus TC#103	ΔT_{103}	3.57	4.17	4.77
ID fan motor case temperature	T_{IDF}	13.75	16.13	17.93
OD fan motor case temperature	T_{ODF}	13.75	23.88	25.07
reversing valve temperature change, discharge side	ΔT_{RVD}	2.38	2.98	3.57
reversing valve temperature change, suction side	ΔT_{RVS}	3.57	4.17	4.77

* Needed a six scan moving window to prevent oscillation around the three standard deviation line.
Underlined entries had sample sizes that oscillated around the three standard deviation line.

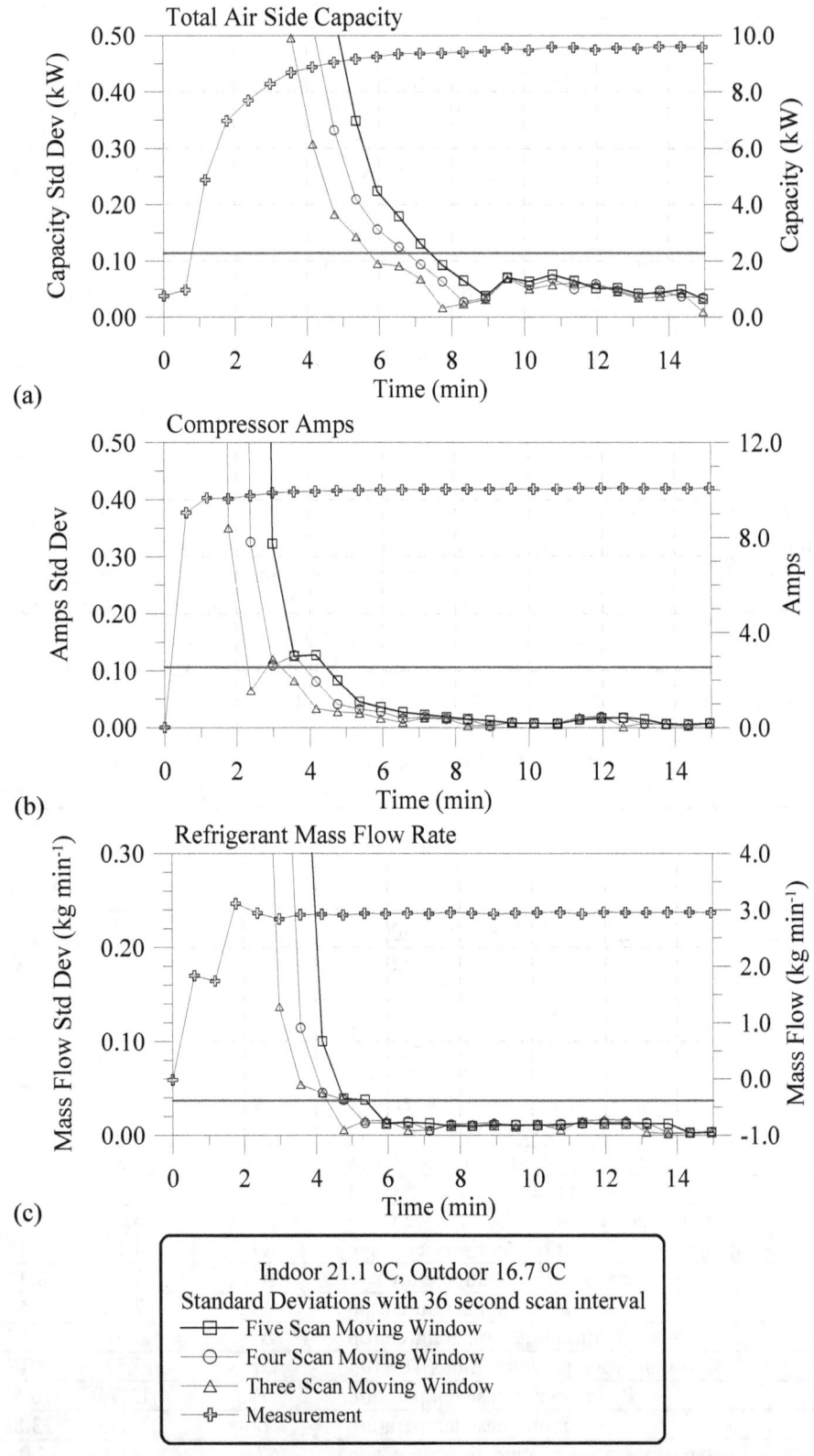

Figure 4.3.1.1. Start-up variations of total air side capacity, compressor amps, refrigerant mass flow rate and their moving window standard deviations for a no-fault test

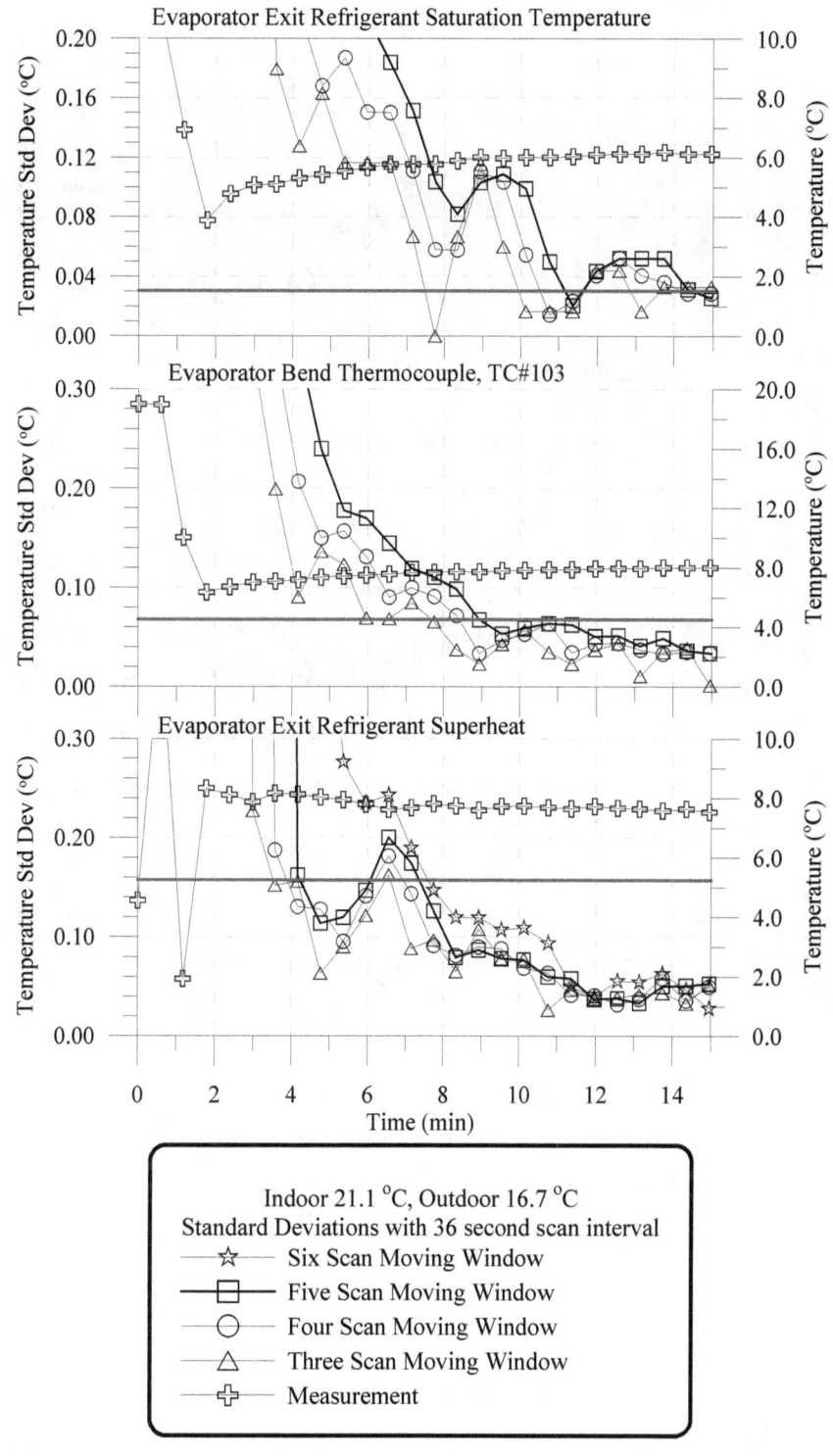

Figure 4.3.1.2. Start-up variation of T_{ER}, T_{E103}, ΔT_{shE} and their moving window standard deviations for a no-fault test

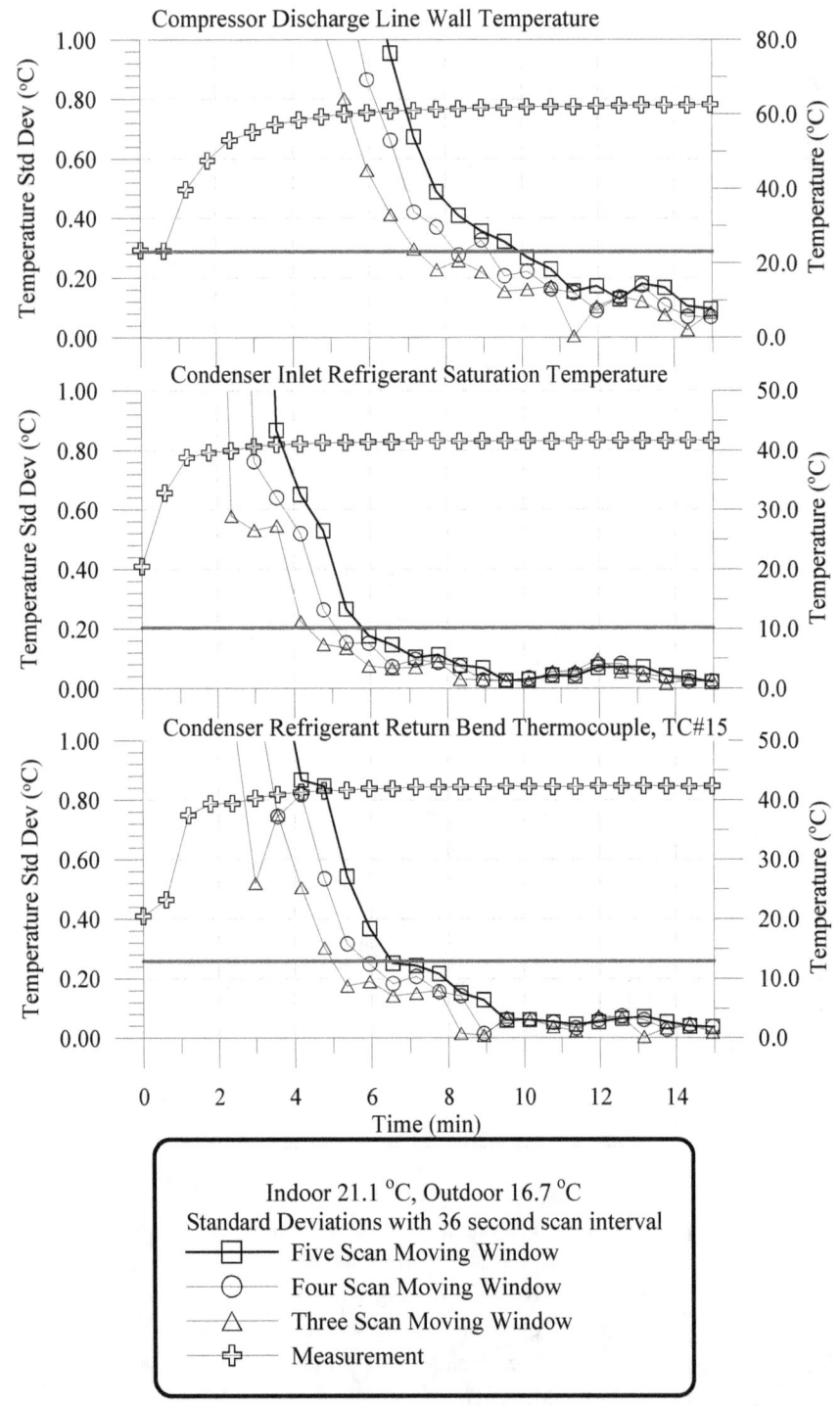

Figure 4.3.1.3. Transient variation of T_D, T_{CR}, and T_{C15} for a no-fault test at an outdoor ambient of 16.7 °C

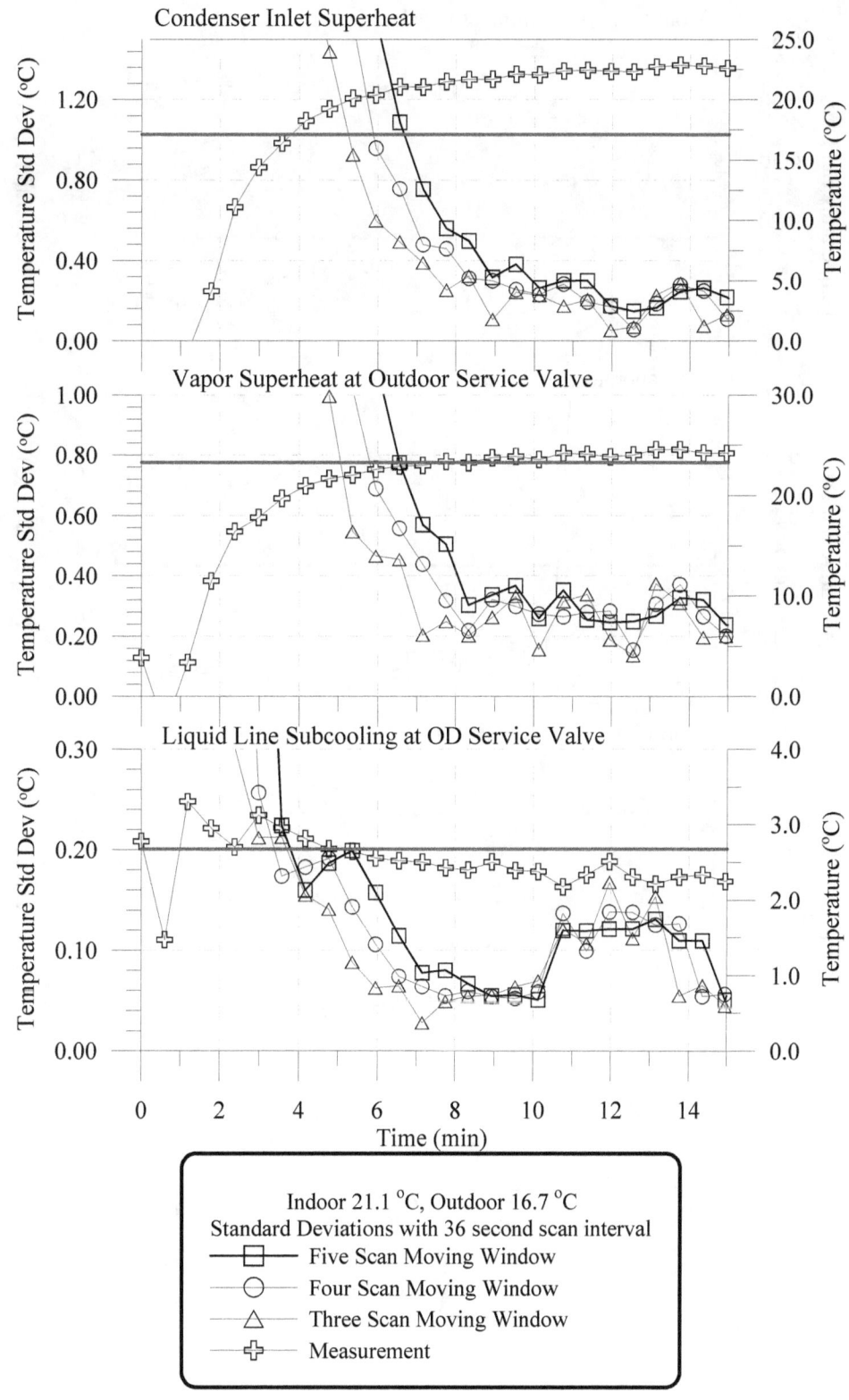

Figure 4.3.1.4. Transient variation of ΔT_{shC}, ΔT_{shV}, and ΔT_{scV} for a no-fault test at an outdoor ambient of 16.7 °C

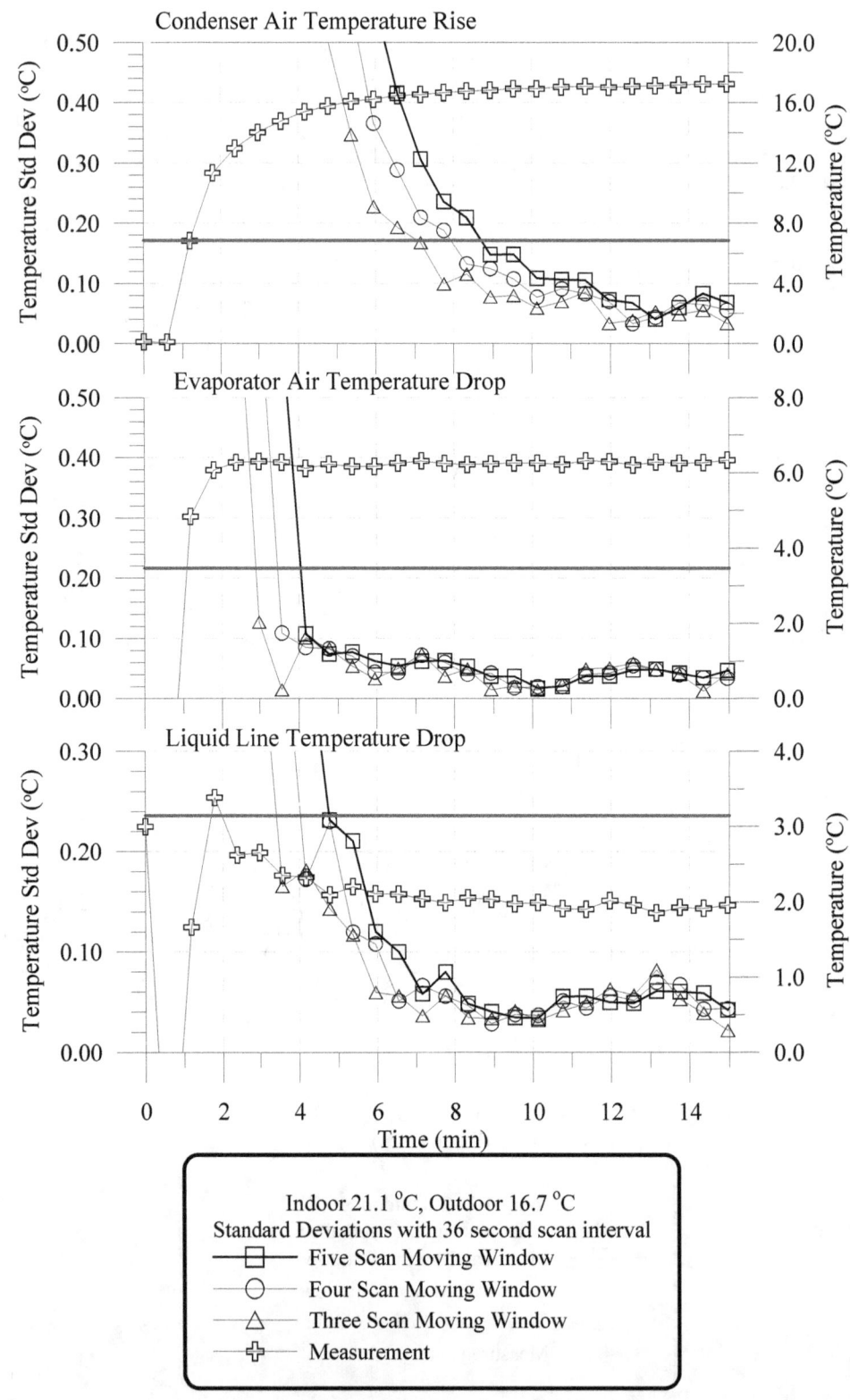

Figure 4.3.1.5. Transient variation of ΔT_{CA}, ΔT_{EA}, and ΔT_{LL} for a no-fault test at an outdoor ambient of 16.7 °C

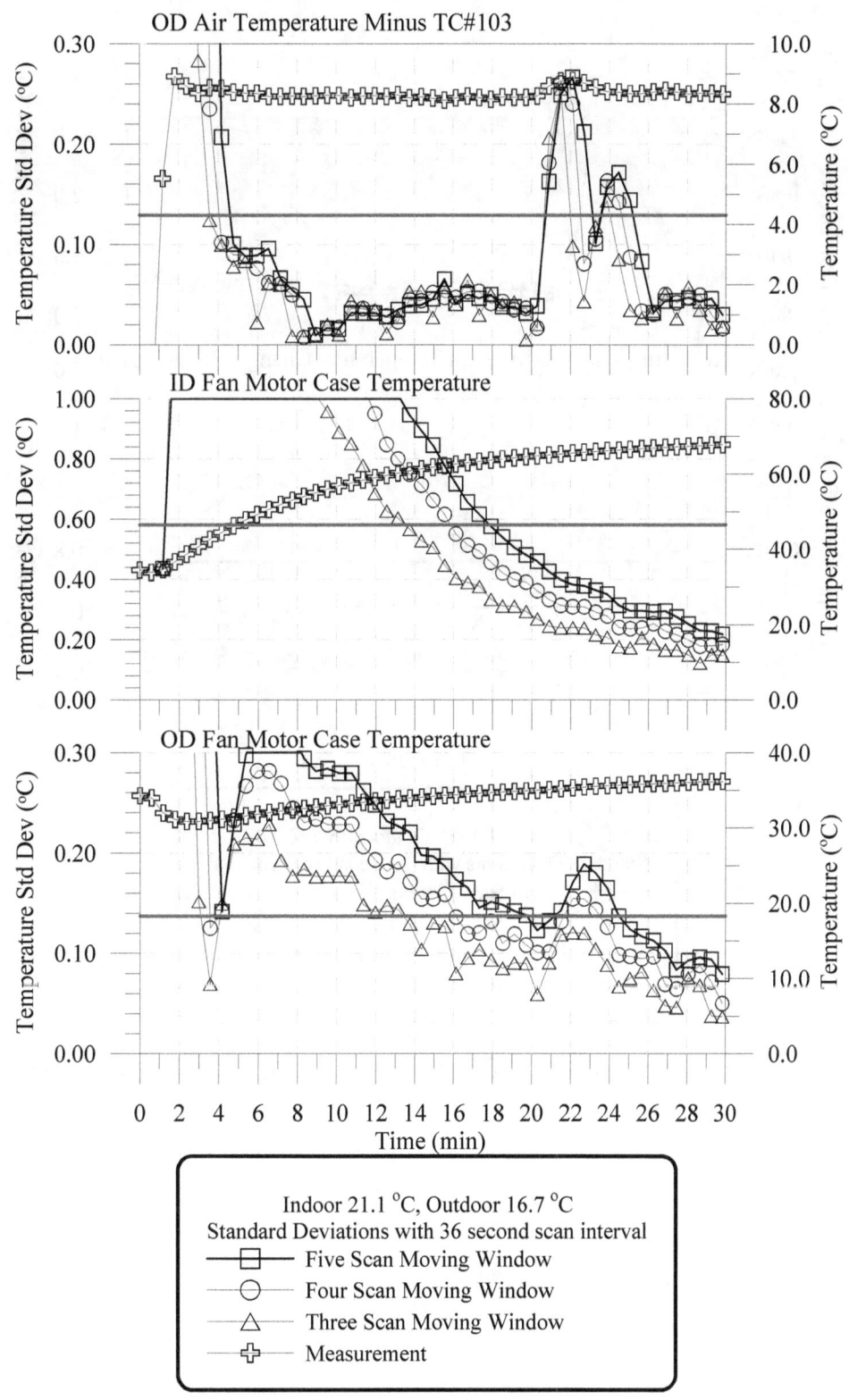

Figure 4.3.1.6. Transient variation of ΔT_{103}, T_{IDF}, and T_{ODF} for a no-fault test at an outdoor ambient of 16.7 °C

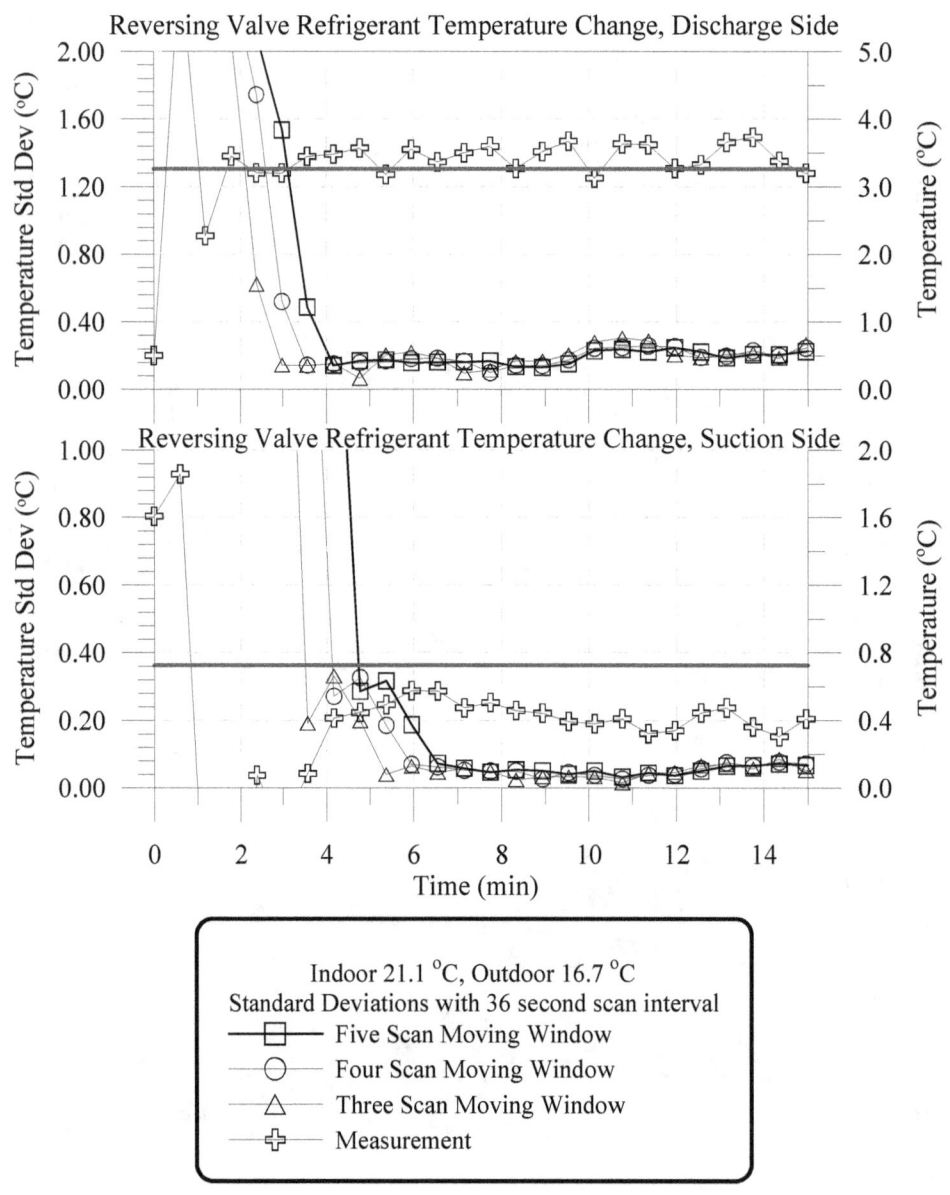

Figure 4.3.1.7. Transient variation of ΔT_{RVD} and ΔT_{RVS} for a no-fault test at an outdoor ambient of 16.7 °C

4.3.2 Start-up variation of features at an indoor/outdoor dry-bulb temperature of 21.1 °C /-8.3 °C

Table 4.3.2.1 presents the time required for the moving window standard deviations of the selected features to remain below a three standard deviation line. Figures 4.3.2.1 through 4.3.2.7 present this information graphically. The solid horizontal line in each figure indicates three standard deviations measured during the fault-free steady-state tests at the given ambient conditions. Underlined entries in Table 4.3.2.1 indicate that the feature's moving window standard deviation, for that moving window scan size, fell below the three standard deviation line and rose above it again before remaining below, i.e. the feature oscillated around the three standard deviation line.

Table 4.3.2.1. Time required for features to remain below the three standard deviation line for various moving window sample sizes at an indoor/outdoor dry-bulb temperature of 21.1/-8.3 °C

Feature Name	Feature Symbol	Three Scans	Four Scans	Five Scans
		minutes		
total air side capacity	Q_{CA}	4.15	4.73	5.33
compressor amps	-	5.92	6.52	7.10
refrigerant mass flow rate	m_R	5.33	5.92	6.52
evaporator exit saturation temperature	T_{ER}	4.15	4.73	5.33
evaporator bend thermocouple, TC#103	T_{E103}	4.15	4.73	5.33
evaporator exit superheat	ΔT_{shE}	4.15	4.73	5.33
compressor discharge wall temperature	T_D	<u>7.70</u>	<u>11.25</u>	12.43*
condenser inlet saturation temperature	T_{CR}	3.55	4.15	4.73
condenser bend thermocouple, TC#15	T_{C15}	3.55	4.15	4.73
condenser inlet superheat	ΔT_{shC}	<u>15.40</u>	16.58	<u>20.13</u>
vapor superheat at outdoor service valve	ΔT_{shV}	<u>11.25</u>	12.43	15.40
liquid line subcooling at outdoor service valve	ΔT_{scV}	4.15	4.73	5.33
condenser air temperature rise	ΔT_{CA}	4.73	5.33	5.92
evaporator air temperature drop	ΔT_{EA}	3.55	4.15	4.73
liquid line temperature drop	ΔT_{LL}	4.73	5.33	5.92
outdoor temperature minus TC#103	ΔT_{103}	4.15	4.73	5.33
ID fan motor case temperature	T_{IDF}	15.40	17.77	20.13
OD fan motor case temperature	T_{ODF}	2.97	3.55	4.15
reversing valve temperature change, discharge side	ΔT_{RVD}	4.73	5.33	5.92
reversing valve temperature change, suction side	ΔT_{RVS}	4.15	4.73	5.33

* Needed a six scan moving window to prevent oscillation around the three standard deviation line.
<u>Underlined</u> entries had sample sizes that oscillated around the three standard deviation line.

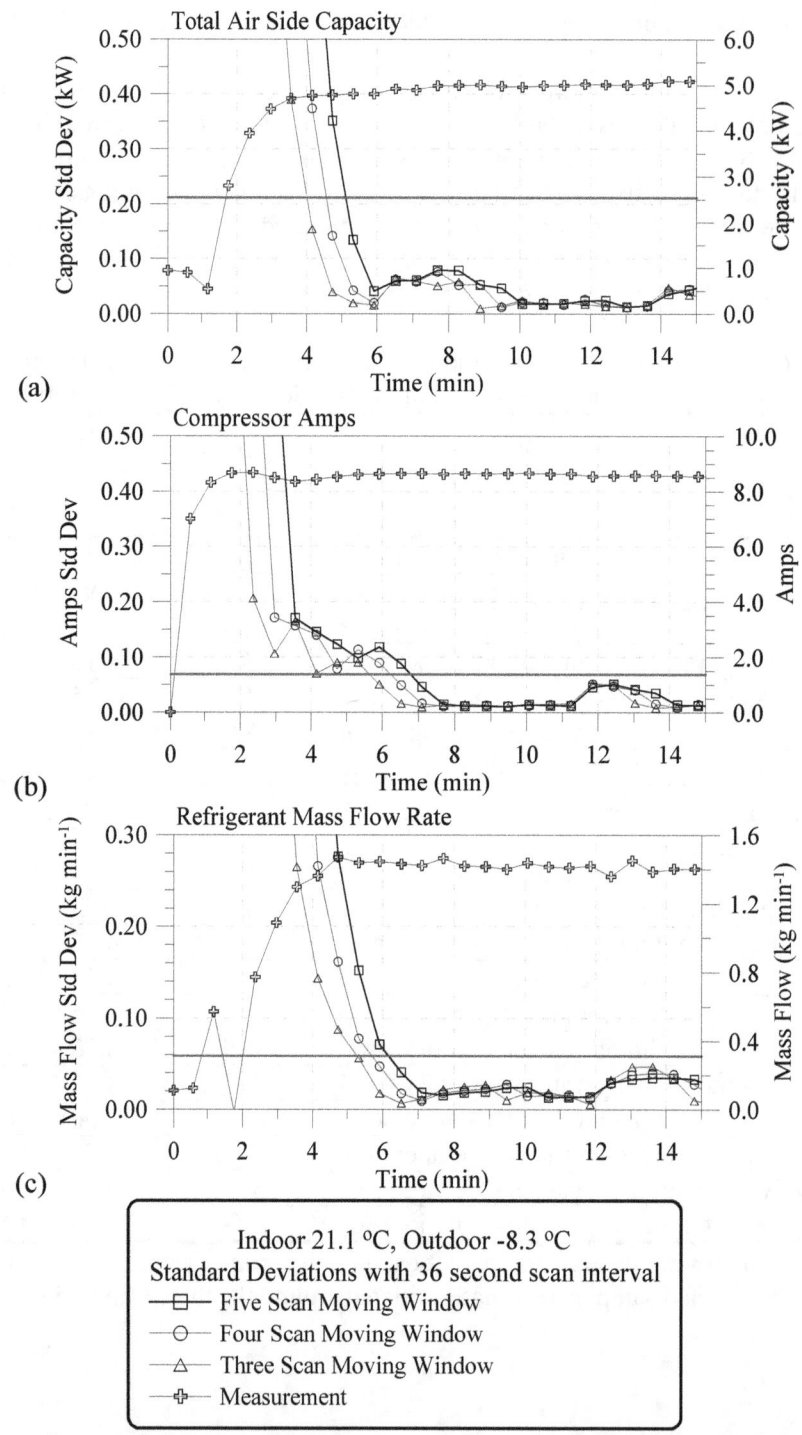

Figure 4.3.2.1. Transient total air side capacity, compressor amps, and refrigerant mass flow rate with moving window standard deviations for a no-fault test at an outdoor ambient of -8.3 °C

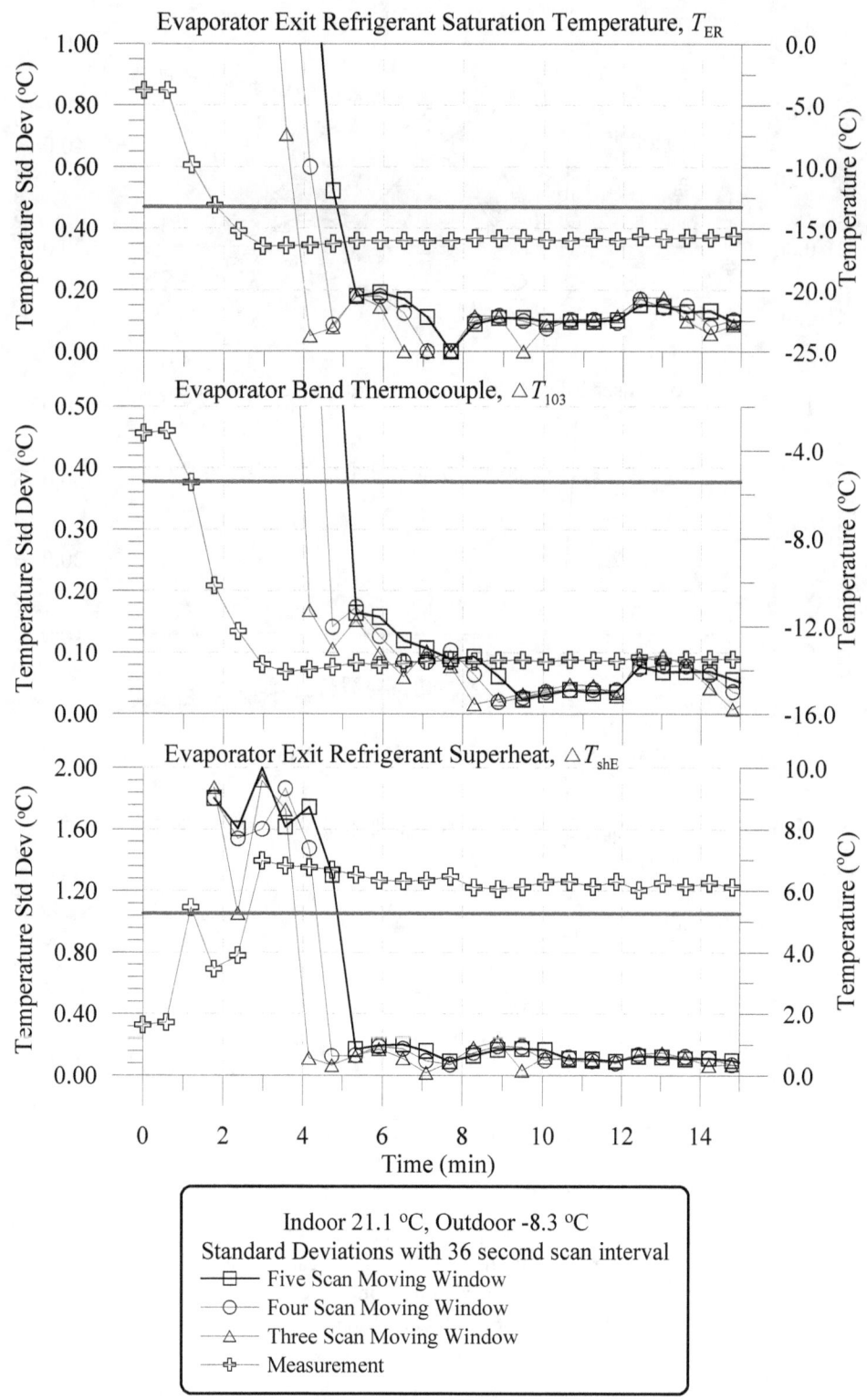

Figure 4.3.2.2. Transient variation of T_{ER}, T_{E103}, and ΔT_{shE} for a no-fault test at an outdoor dry-bulb temperature of -8.3 °C

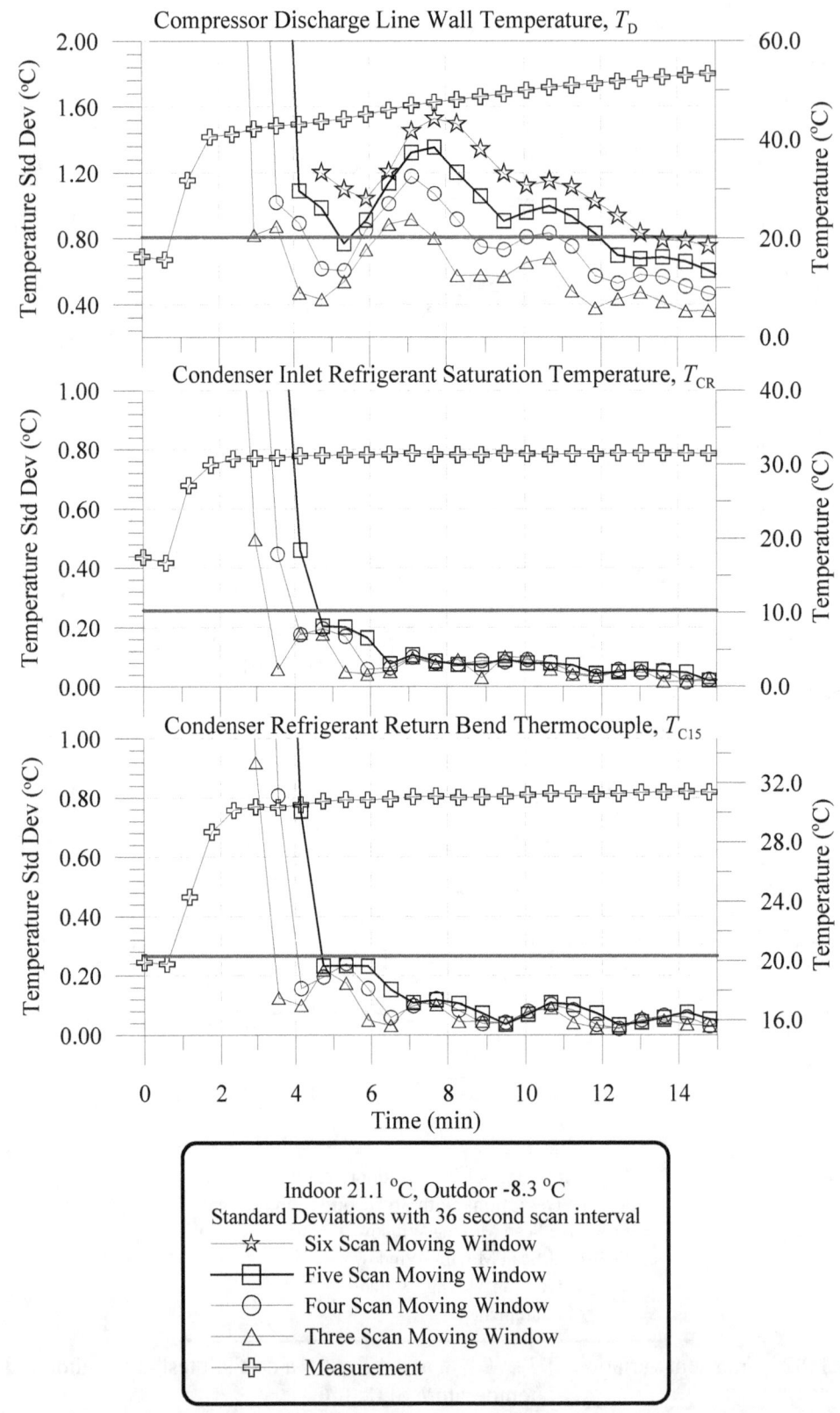

Figure 4.3.2.3. Transient variation of T_D, T_{CR}, and T_{C15} for a no-fault test at an outdoor dry-bulb temperature of -8.3 °C

48

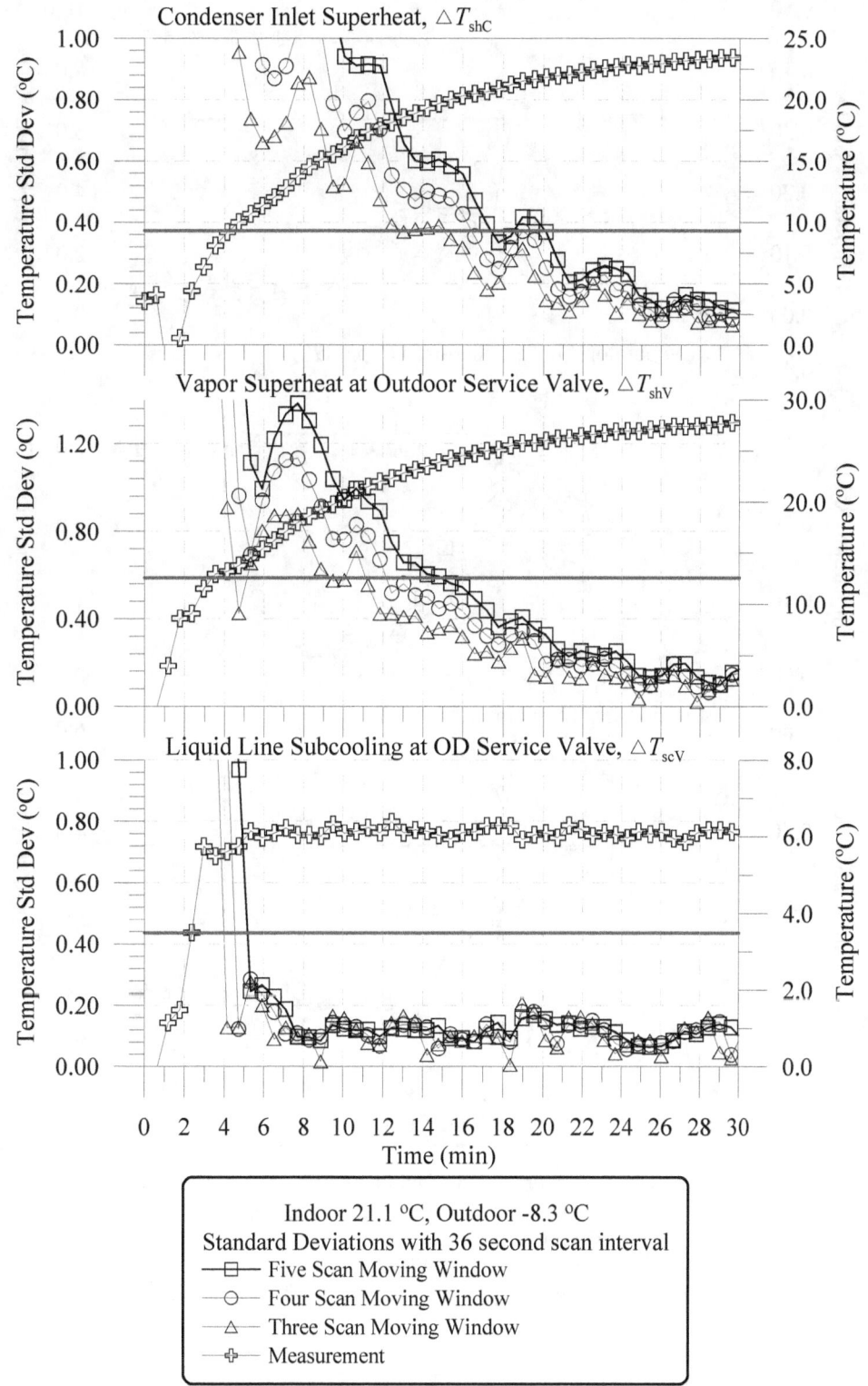

Figure 4.3.2.4. Transient variation of ΔT_{shC}, ΔT_{shV}, and ΔT_{scV} for a no-fault test at an outdoor dry-bulb temperature of -8.3 °C

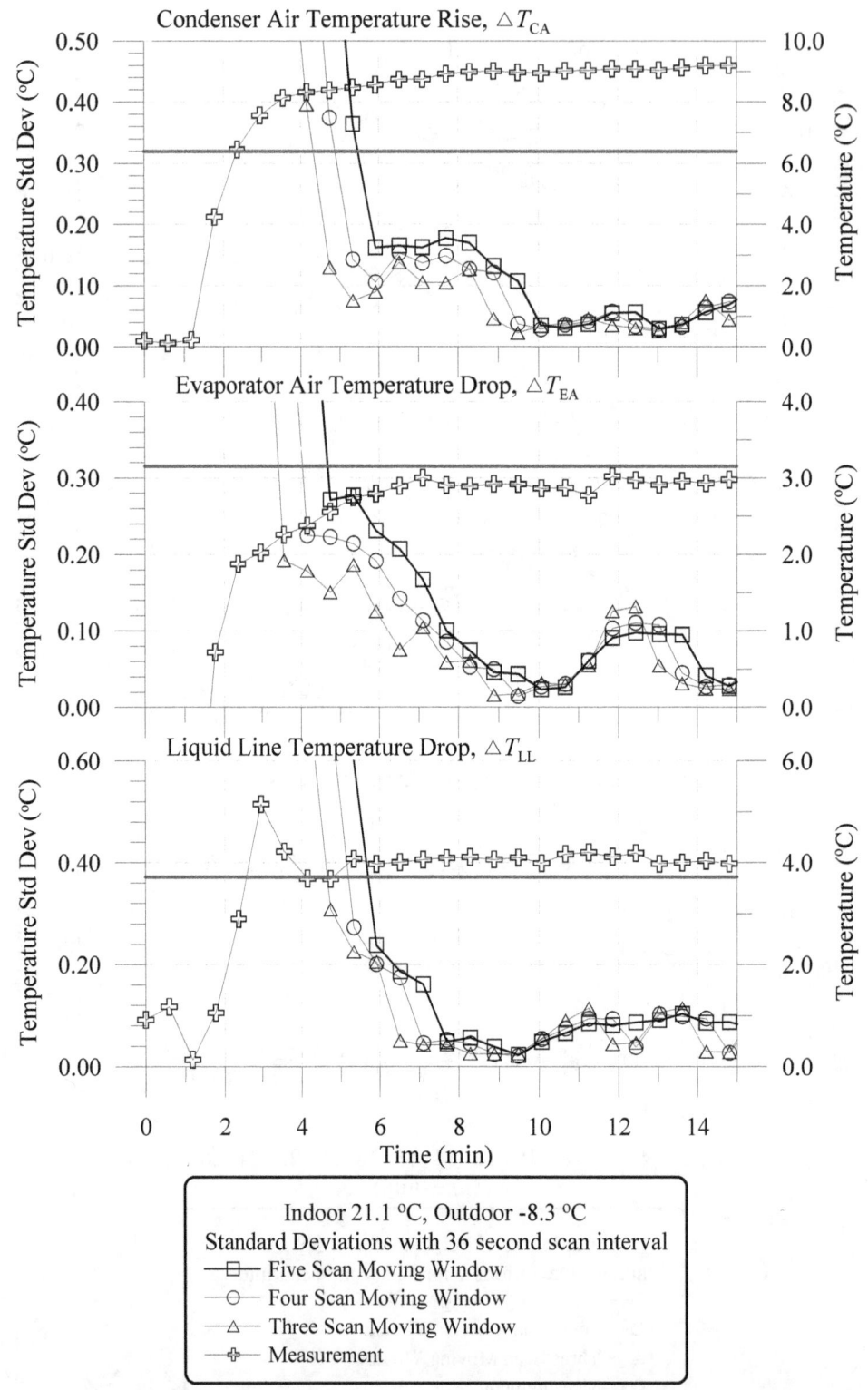

Figure 4.3.2.5. Transient variation of ΔT_{CA}, ΔT_{EA}, and ΔT_{LL} for a no-fault test at an outdoor dry-bulb temperature of -8.3 °C

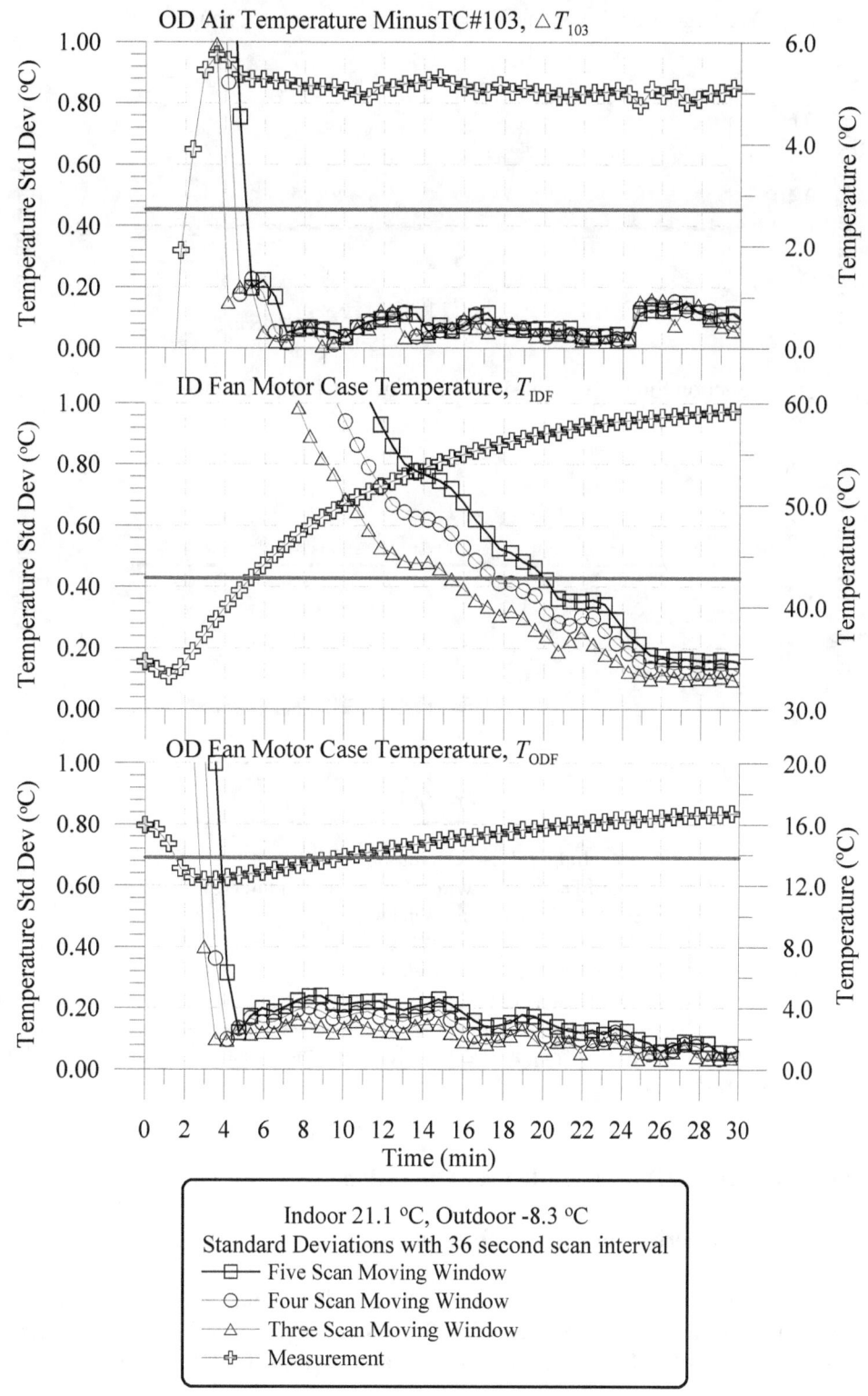

Figure 4.3.2.6. Transient variation of ΔT_{103}, T_{IDF}, and T_{ODF} for a no-fault test at an outdoor dry-bulb temperature of -8.3 °C

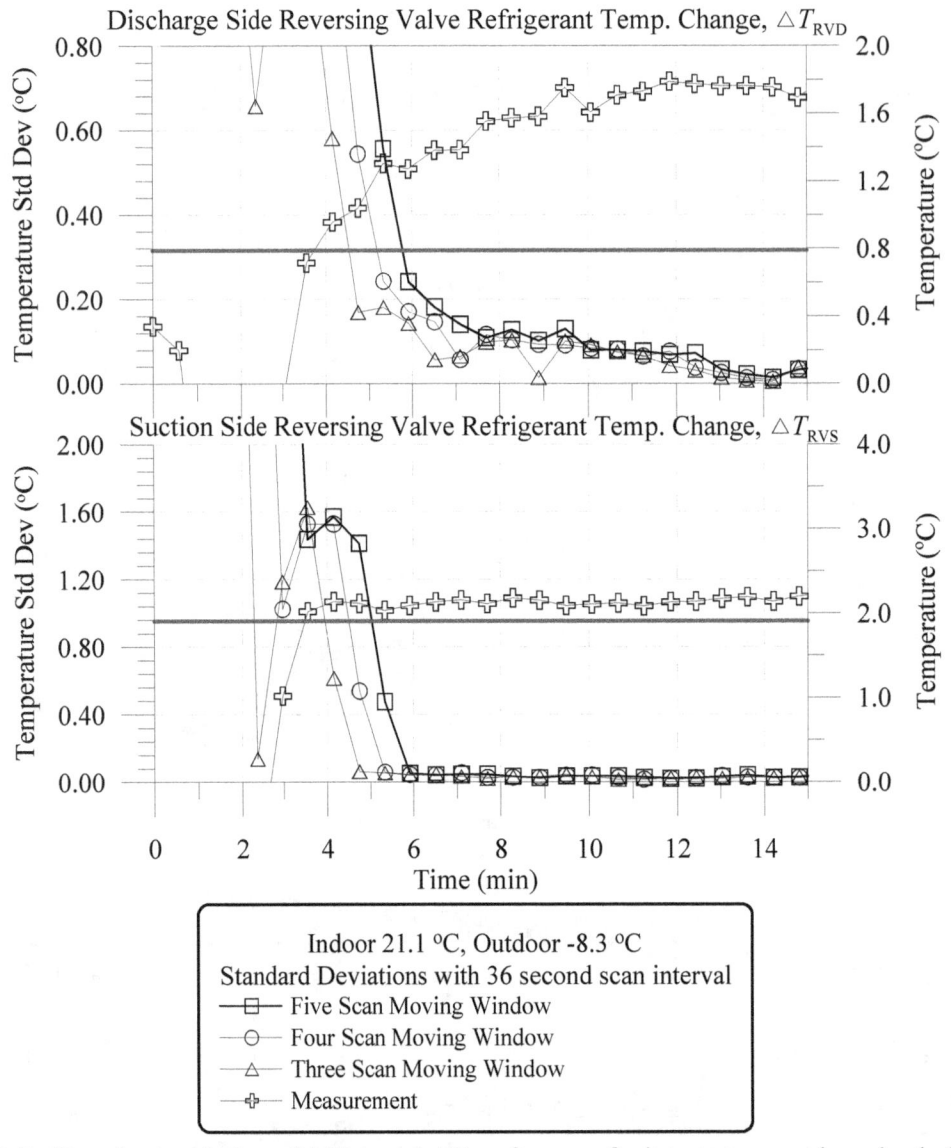

Figure 4.3.2.7. Transient variation of ΔT_{RVD} and ΔT_{RVS} for a no-fault test at an outdoor dry-bulb
temperature of -8.3 °C

4.4 Frosting Characteristics During Fault-Free Operation

4.4.1 System performance during outdoor coil frosting tests

During the frosting tests, the average outdoor dry-bulb and dew-point temperatures plus or minus one
standard deviation were (2.2±0.6) °C and (-2.7±0.8) °C (68 % RH), respectively, and the indoor dry-bulb
temperature was (21.1±0.2) °C. The presence of frost manifested itself in an increased air pressure drop,
as shown in Figure 4.4.1.

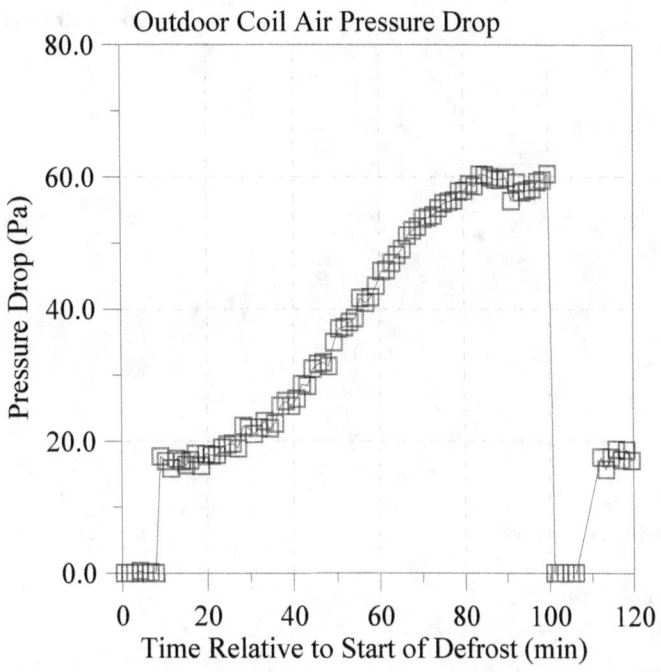

Figure 4.4.1. Outdoor coil air pressure drop during frosting at an outdoor dry-bulb/dew-point temperature of (2.2±0.6) °C/(-2.7±0.8) °C and indoor dry-bulb temperature of (21.1±0.2) °C

Figure 4.4.1.1 shows the air side heating capacity, refrigerant side capacity, COP, compressor power, and compressor amps as a function of time from the last defrost initiation.

53

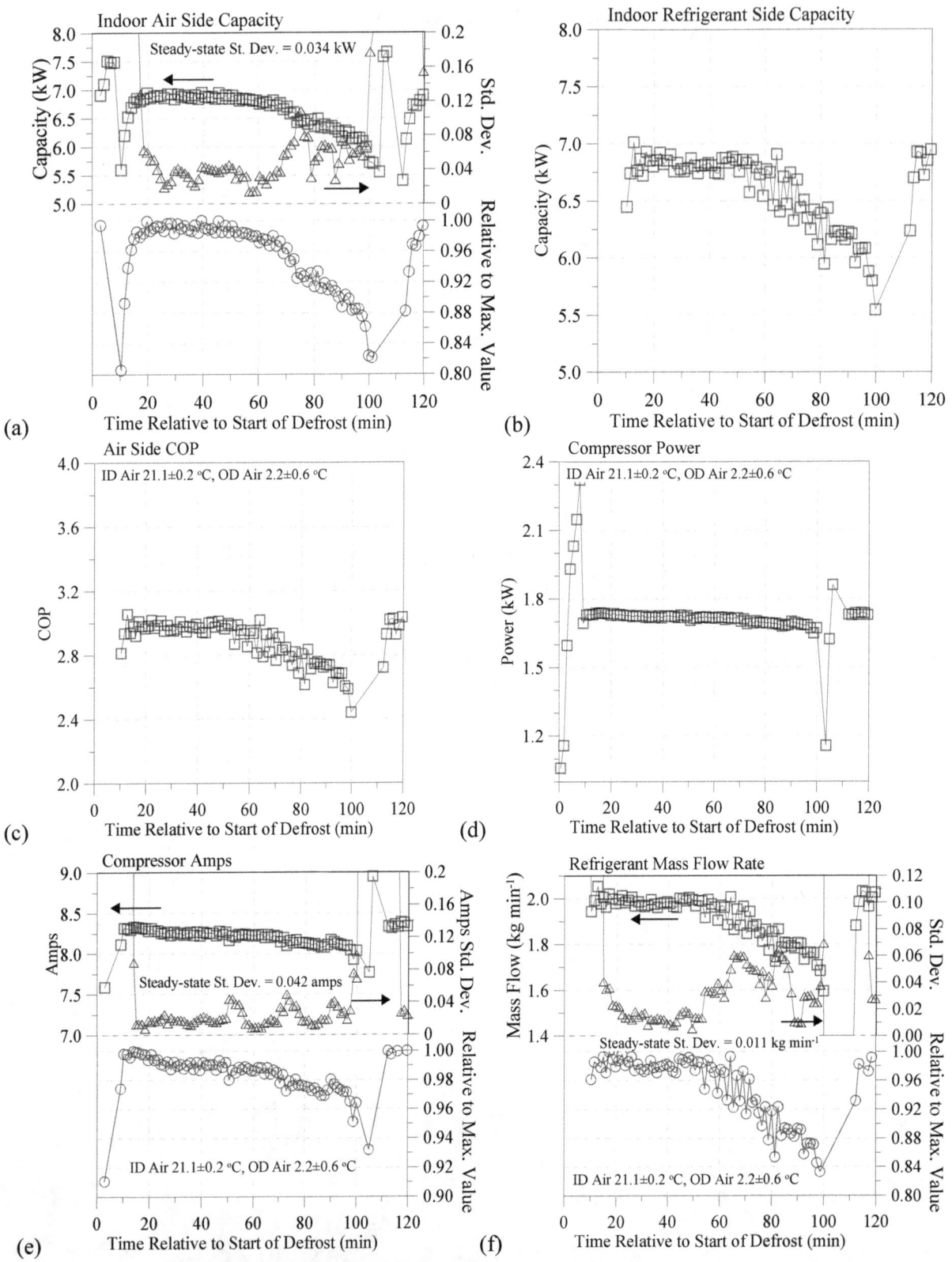

Figure 4.4.1.1. (a) Air side heating capacity, (b) refrigerant side heating capacity, (c) COP, (d) compressor power, (e) compressor amps, and (f) refrigerant mass flow rate for a frosting no-fault test

4.4.2 Feature variations during outdoor coil frosting

Figures 4.4.2.1 through 4.4.2.5 show selected feature's variations during a frosting test. Figure 4.4.2.2 shows that condenser inlet refrigerant superheat starts to fluctuate approximately 60 minutes after defrost starts. This may indicate the presence of two-phase refrigerant in the vapor line. This fluctuation also appears in Figure 4.4.2.5 on the suction side of the reversing valve; again two-phase refrigerant may be the culprit, but no direct observations were made.

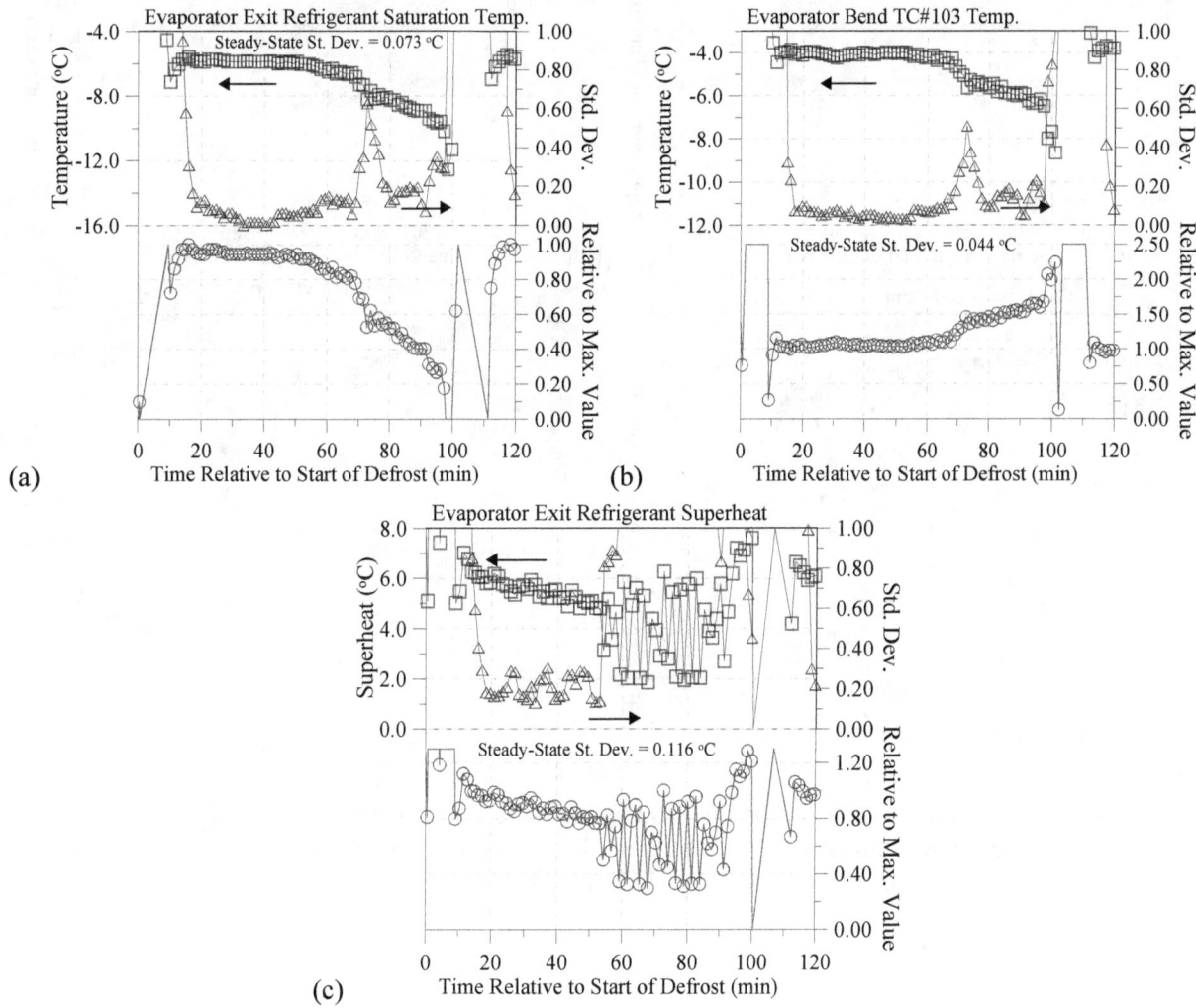

Figure 4.4.2.1. (a) Evaporator exit saturation temperature, (b) evaporator bend thermocouple #103, and (c) evaporator exit superheat for a frosting no-fault test

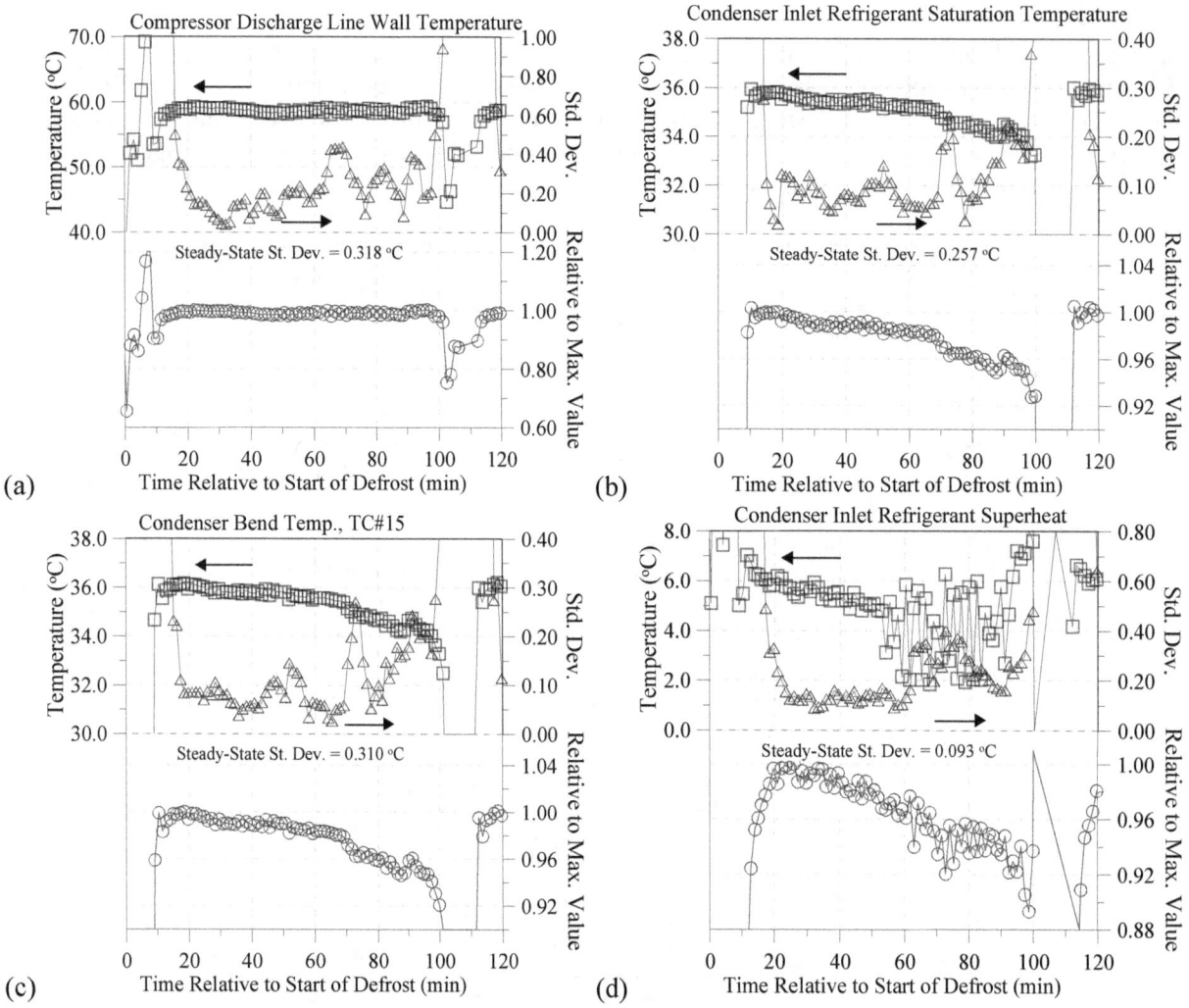

Figure 4.4.2.2. (a) Compressor discharge line wall temperature, (b) condenser inlet refrigerant vapor saturation temperature, (c) condenser bend thermocouple, T_{C15}, and (d) condenser inlet refrigerant superheat for frosting no-fault tests

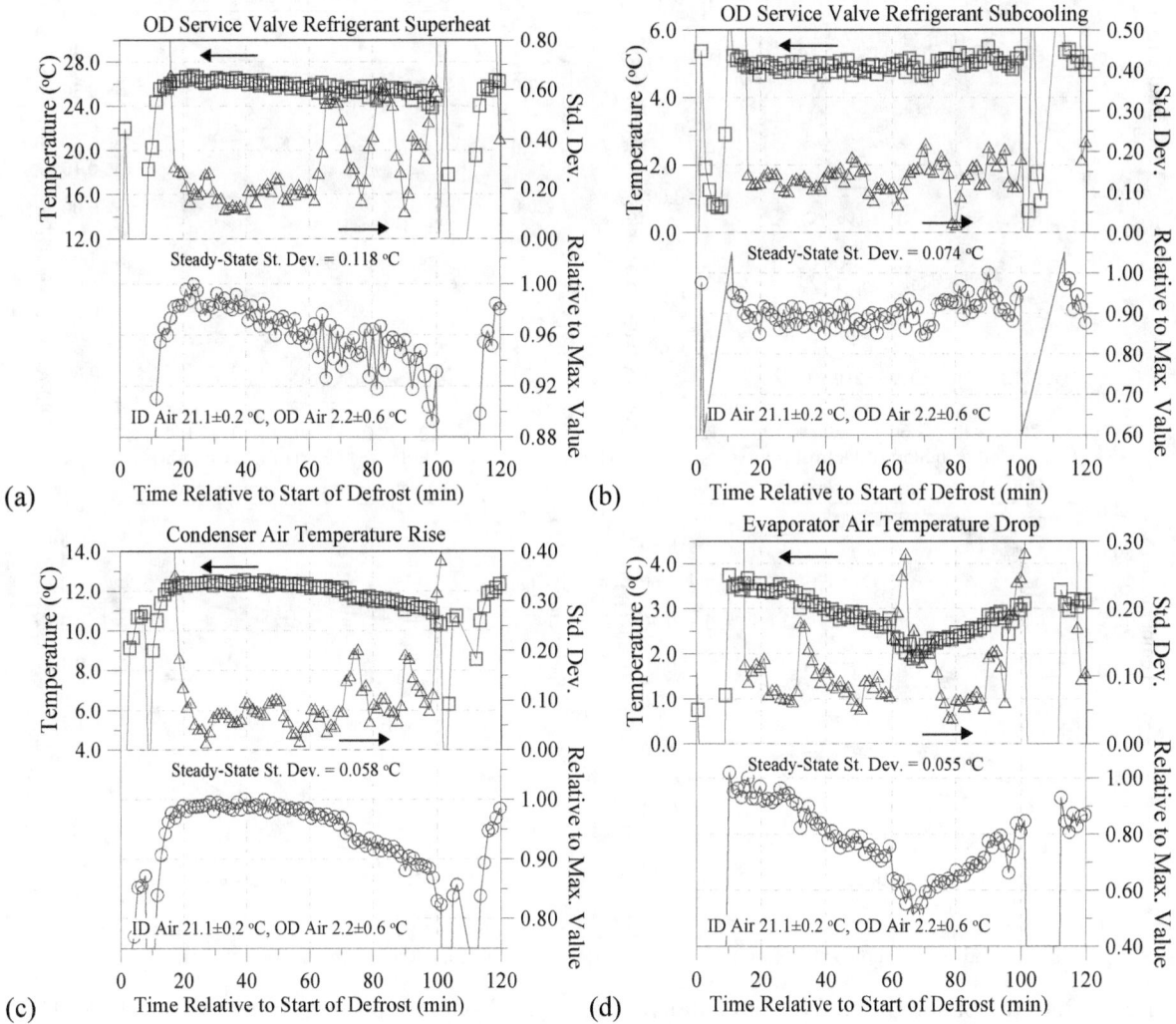

Figure 4.4.2.3. (a) Refrigerant vapor superheat at the outdoor service valve, (b) refrigerant liquid subcooling at the outdoor service valve, (c) condenser air temperature rise, and (d) evaporator air temperature drop for frosting no-fault test

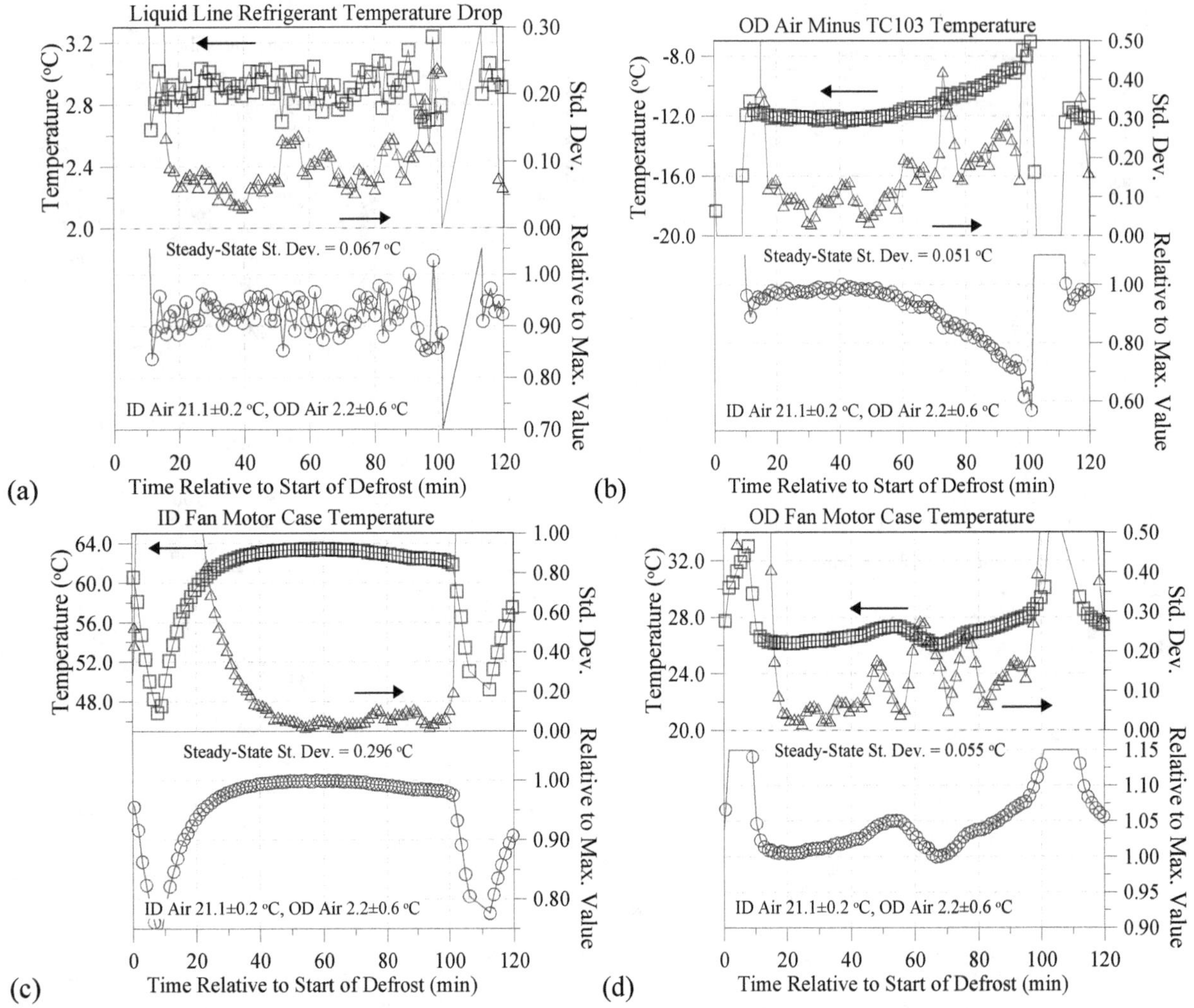

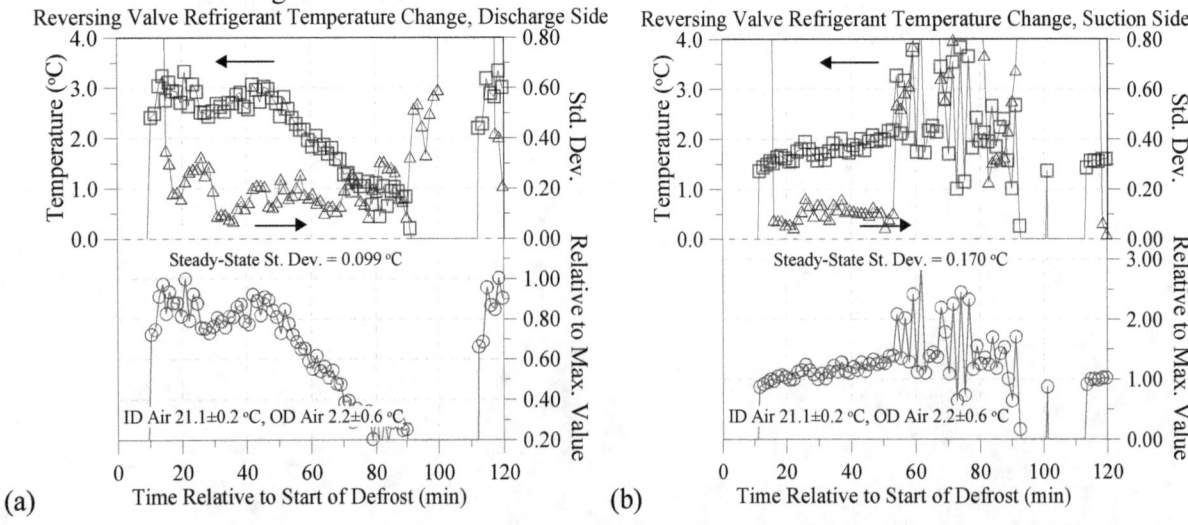

Figure 4.4.2.4. (a) Liquid line refrigerant temperature drop, (b) outdoor air temperature minus TC#103, (c) indoor fan motor case temperature, and (d) outdoor fan motor case temperature for frosting no-fault test

Figure 4.4.2.5. (a) Reversing valve refrigerant temperature change on the discharge side and (b) reversing valve refrigerant temperature change on the suction side for frosting no-fault test

4.5 Fault-Free Test Repeatability

Repeatability tests were performed to determine the variability in feature values at the same nominal test conditions (indoor dry-bulb temperature and outdoor dry-bulb temperature-dry coil) due to the inability to exactly replicate test conditions. When performing tests at the same nominal test conditions, the actual dry-bulb temperatures can never be exactly the same from one test to the next. This becomes a concern when a feature's value at a given test condition is needed to calculate a residual during testing at the same environmental conditions with a fault applied to the system. This means that the inability to exactly match test conditions from one test to the next introduces an uncertainty in the resulting comparisons because the same test conditions cannot be exactly repeated.

The residual is calculated as the difference between the current feature value and the reference value as shown in Equation 4.5.1.

$$R = F - NF \tag{4.5.1}$$

Where: R = residual
           F = fault imposed value of feature
           NF = fault-free value of feature

The reference values listed in the preceding tables are the arithmetic averages of several tests during fault-free steady-state testing at the respective ambient conditions. Repeatability tests revealed the variation in these feature values over time and from one test condition to the next. The "nominal" test conditions are those conditions which are being repeated, which are also the reference case test conditions. The tables presented below indicate percent changes from reference state values; these percent values for temperature residuals should not be used to draw any conclusions because many times the residual (or temperature change relative to the reference value) is too small to be measured with reasonable uncertainty. The most important parameter in the tables is the residual.

Table 4.5.1 compares selected reference features and the average of the four repeat tests (Tests #1, #2, #3, and #4) for a given nominal test condition of 21.1 °C indoor and -8.3 °C outdoor dry-bulb temperatures. The highest percent difference between the reference and the repeat feature values occurred for ΔT_{RVS} and ΔT_{scV}. This is just one example of the variation in system features when trying to exactly replicate test conditions from one test in another test. Figures 4.5.1 and 4.5.2 graphically show the test temperature conditions listed as an average value plus or minus one standard deviation in these figures.

Table 4.5.1. Variation in features due to variation in ambient condition repeatability for a given nominal test condition of 21.1 °C indoor and -8.3 °C outdoor dry-bulb temperatures

Feature Name	Feature Symbol	Reference Value	Repeat Test Value	Residual	% Difference wrt the Reference
total air side capacity (kW)	Q_{CA}	5.15±0.07	5.11±0.04	-0.04	-0.78
compressor power (kW)	W_{comp}	1.604±0.004	1.607±0.002	0.003	1.8e-5
refrigerant mass flow rate (kg min^{-1})	m_R	1.386±0.020	1.381±0.025	-0.005	-0.36
coefficient of performance	COP	2.275±0.035	2.264±0.042	-0.011	-0.48
indoor unit refrigerant side capacity (kW)	Q_{CR}	4.937±0.007	4.919±0.091	-0.018	-0.36
evaporator exit saturation temperature (°C)	T_{ER}	-15.47±0.16	-15.43±0.10	0.04	0.26
evaporator bend thermocouple, TC#103 (°C)	T_{E103}	-13.20±0.13	-13.20±0.08	0.00	0.00
evaporator exit superheat (°C)	ΔT_{shE}	5.76±0.35	5.93±0.27	0.17	2.95
compressor discharge wall temperature (°C)	T_D	59.51±0.27	59.83±0.12	0.32	0.54
condenser inlet saturation temperature (°C)	T_{CR}	31.54±0.09	31.66±0.06	0.12	0.38
condenser bend thermocouple, TC#15 (°C)	T_{C15}	31.81±0.09	31.97±0.07	0.16	0.50
condenser inlet superheat (°C)	ΔT_{shC}	25.33±0.13	25.64±0.12	0.31	1.22
vapor superheat at outdoor service valve (°C)	ΔT_{shV}	29.99±0.20	30.22±0.14	0.23	0.77
liquid line subcooling at outdoor service valve (°C)	ΔT_{scV}	6.49±0.15	6.19±0.11	-0.30	-4.62
condenser air temperature rise (°C)	ΔT_{CA}	9.34±0.11	9.35±0.08	0.01	0.11
evaporator air temperature drop (°C)	ΔT_{EA}	2.75±0.08	2.71±0.10	-0.04	-1.45
liquid line temperature drop (°C)	ΔT_{LL}	4.19±0.12	3.92±0.09	-0.27	-6.44
outdoor temperature minus TC#103 (°C)	ΔT_{103}	4.81±0.15	4.86±0.12	0.05	1.04
ID fan motor case temperature (°C)	T_{IDF}	59.92±0.14	60.19±0.05	0.27	0.45
OD fan motor case temperature (°C)	T_{ODF}	17.59±0.23	17.75±0.03	0.16	0.91
reversing valve temperature change, discharge side (°C)	ΔT_{RVD}	1.81±0.11	1.82±0.07	0.01	0.55
reversing valve temperature change, suction side (°C)	ΔT_{RVS}	2.08±0.32	2.34±0.15	0.26	12.50

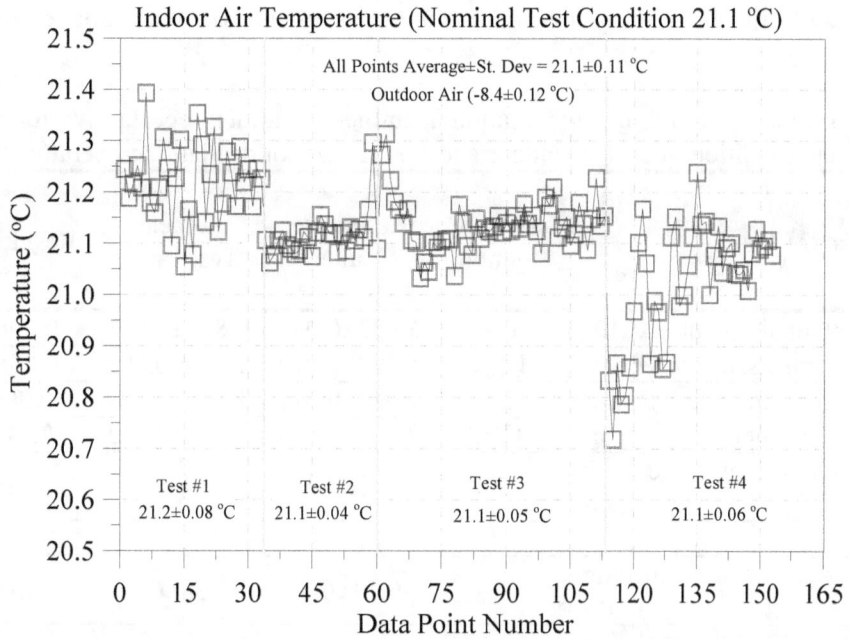

Figure 4.5.1. Variation of indoor air dry-bulb temperature for several test cases with nominal conditions of 21.1 °C indoor and -8.3 °C outdoor dry-bulb temperatures

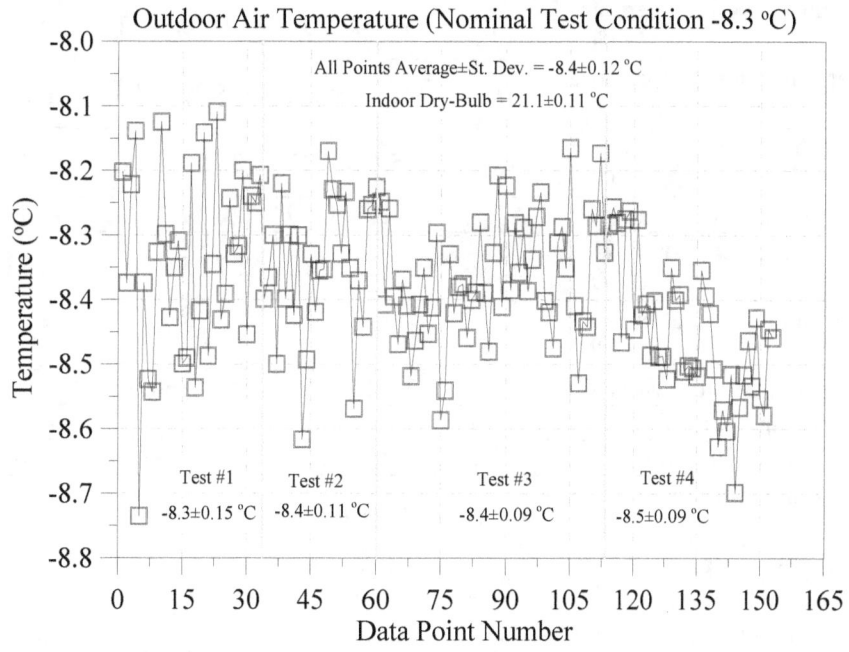

Figure 4.5.2. Variation of outdoor air dry-bulb temperature for several test cases with nominal conditions of 21.1 °C indoor and -8.3 °C outdoor dry-bulb temperatures

Table 4.5.2 compares selected features of the reference and a repeat tests (Tests #1 and #2) for a given nominal test condition of 21.1 °C indoor and 1.7 °C outdoor dry-bulb temperatures. The highest percent difference between the reference and the repeat feature values occurred for ΔT_{RVD} and ΔT_{shE}. Figures 4.5.3 and 4.5.4 graphically show the test temperature conditions listed as an average value plus or minus one standard deviation in these figures. As Figure 4.5.3 shows, test conditions with the same average values do not necessarily have the same temperature profile, nor will they produce exactly the same feature values.

Table 4.5.2. Variation in features due to variation in ambient condition repeatability for a given nominal test condition of 21.1 °C indoor and 1.7 °C outdoor dry-bulb temperatures

Feature Name	Feature Symbol	Reference Value	Repeat Test Value	Residual	% Difference wrt the Reference
total air side capacity (kW)	Q_{CA}	6.822±0.034	6.828±0.035	0.006	0.09
compressor power (kW)	W_{comp}	1.708±0.012	1.716±0.002	0.008	0.47
refrigerant mass flow rate (kg min^{-1})	m_R	1.949±0.011	1.940±0.009	-0.009	-0.45
coefficient of performance	COP	2.970±0.024	2.956±0.015	-0.014	-0.49
indoor unit refrigerant side capacity (kW)	Q_{CR}	6.741±0.037	6.726±0.033	-0.015	-0.22
evaporator exit saturation temperature (°C)	T_{ER}	-6.34±0.07	-6.38±0.05	-0.04	0.60
evaporator bend thermocouple, TC#103 (°C)	T_{E103}	-4.42±0.05	-4.48±0.02	-0.06	1.30
evaporator exit superheat (°C)	ΔT_{shE}	6.31±0.12	6.72±0.06	0.41	6.47
compressor discharge wall temperature (°C)	T_D	59.04±0.32	59.80±0.11	0.76	1.29
condenser inlet saturation temperature (°C)	T_{CR}	35.22±0.26	35.29±0.06	0.07	0.19
condenser bend thermocouple, TC#15 (°C)	T_{C15}	35.63±0.31	35.90±0.06	0.27	0.77
condenser inlet superheat (°C)	ΔT_{shC}	24.14±0.09	24.59±0.09	0.45	1.87
vapor superheat at outdoor service valve (°C)	ΔT_{shV}	27.22±0.12	27.55±0.12	0.33	1.20
liquid line subcooling at outdoor service valve (°C)	ΔT_{scV}	5.00±0.07	5.08±0.08	0.08	1.68
condenser air temperature rise (°C)	ΔT_{CA}	12.27±0.06	12.45±0.06	0.18	1.47
evaporator air temperature drop (°C)	ΔT_{EA}	3.93±0.05	3.88±0.04	-0.05	-1.16
liquid line temperature drop (°C)	ΔT_{LL}	2.92±0.07	2.85±0.05	-0.08	-2.65
outdoor temperature minus TC#103 (°C)	ΔT_{103}	6.01±0.05	6.21±0.05	0.20	3.41
ID fan motor case temperature (°C)	T_{IDF}	62.95±0.30	63.77±0.05	0.83	1.31
OD fan motor case temperature (°C)	T_{ODF}	24.96±0.05	25.10±0.04	0.14	0.56
reversing valve temperature change, discharge side (°C)	ΔT_{RVD}	3.17±0.10	2.86±0.09	-0.31	-9.79
reversing valve temperature change, suction side (°C)	ΔT_{RVS}	1.08±0.17	1.04±0.09	-0.04	-3.88

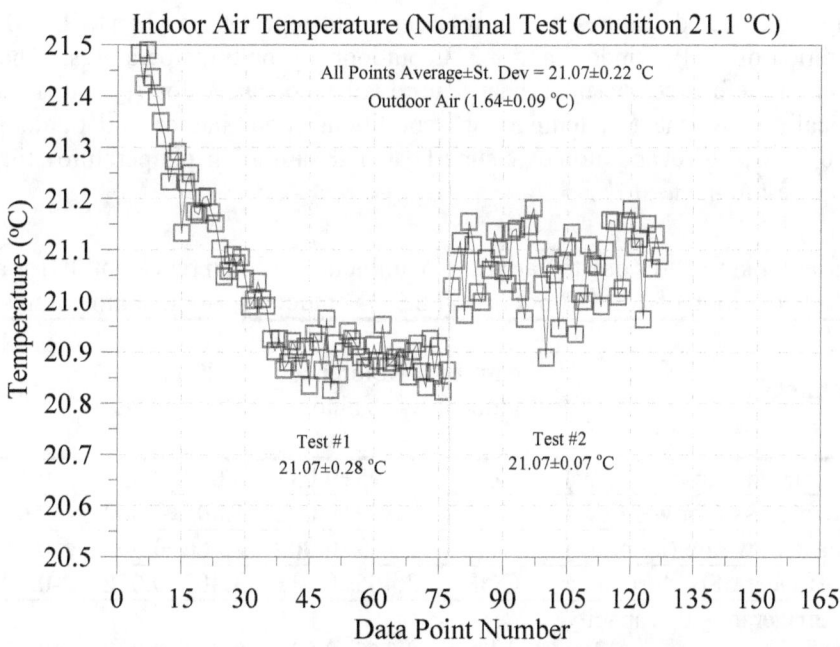

Figure 4.5.3. Variation of indoor air dry-bulb temperature for several test cases with nominal conditions of 21.1 °C indoor and 1.67 °C outdoor dry-bulb temperatures

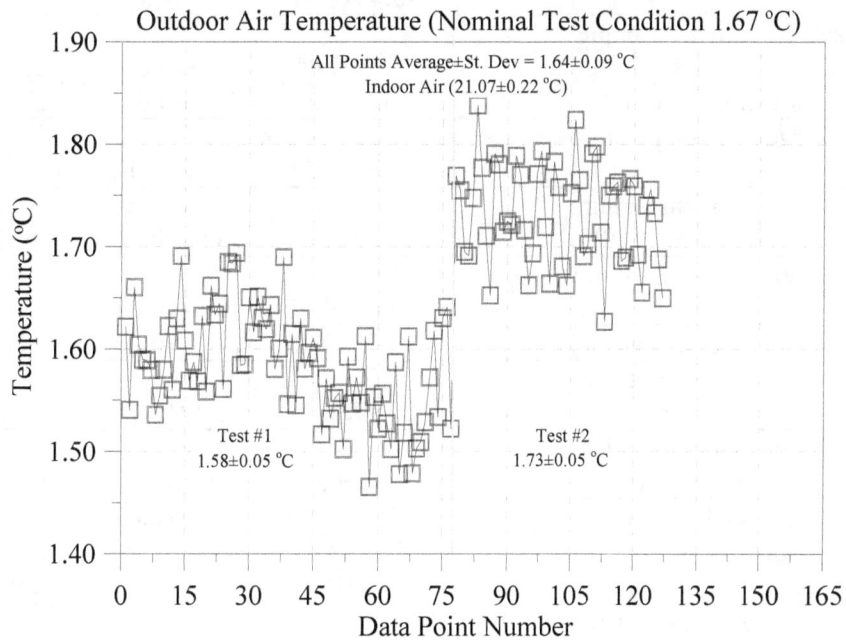

Figure 4.5.4. Variation of outdoor air dry-bulb temperature for several test cases with nominal conditions of 21.1 °C indoor and 1.67 °C outdoor dry-bulb temperatures

Table 4.5.3 compares selected features of the reference and repeat tests (Tests #1 and #2) for a given nominal test condition of 21.1 °C indoor and 8.3 °C outdoor dry-bulb temperatures. The highest percent difference between the reference and the repeat feature values occurred for T_{ER} and ΔT_{shE}. Figures 4.5.5 and 4.5.6 graphically show the test temperature conditions; both indoor and outdoor conditions are matched within 0.1 °C, but evaporator exit superheat and saturation temperature still had substantial variations from one test to the next.

Table 4.5.3. Variation in features due to variation in ambient condition repeatability for a given nominal test condition of 21.1 °C indoor and 8.3 °C outdoor dry-bulb temperatures

Feature Name	Feature Symbol	Reference Value	Repeat Test Value	Residual	% Difference wrt the Reference
total air side capacity (kW)	Q_{CA}	8.195±0.031	8.072±0.031	-0.123	-1.50
compressor power (kW)	W_{comp}	1.809±0.002	1.803±0.002	-0.006	-0.35
refrigerant mass flow rate (kg min^{-1})	m_R	2.379±0.009	2.366±0.008	-0.014	-0.58
coefficient of performance	COP	3.405±0.013	3.402±0.012	-0.003	-0.08
indoor unit refrigerant side capacity (kW)	Q_{CR}	8.045±0.030	8.013±0.028	-0.032	-0.39
evaporator exit saturation temperature (°C)	T_{ER}	-0.44±0.01	-0.63±0.05	-0.189	42.67
evaporator bend thermocouple, TC#103 (°C)	T_{E103}	1.33±0.04	1.18±0.02	-0.153	-11.46
evaporator exit superheat (°C)	ΔT_{shE}	6.69±0.07	7.07±0.06	0.382	5.71
compressor discharge wall temperature (°C)	T_D	60.83±0.11	60.92±0.07	0.091	0.15
condenser inlet saturation temperature (°C)	T_{CR}	38.26±0.04	38.08±0.03	-0.182	-0.47
condenser bend thermocouple, TC#15 (°C)	T_{C15}	39.04±0.05	38.84±0.04	-0.202	-0.52
condenser inlet superheat (°C)	ΔT_{shC}	23.78±0.08	24.13±0.06	0.354	1.49
vapor superheat at outdoor service valve (°C)	ΔT_{shV}	25.97±0.14	26.39±0.14	0.411	1.58
liquid line subcooling at outdoor service valve (°C)	ΔT_{scV}	4.07±0.07	4.05±0.05	-0.019	-0.48
condenser air temperature rise (°C)	ΔT_{CA}	14.91±0.04	14.72±0.03	-0.186	-1.25
evaporator air temperature drop (°C)	ΔT_{EA}	4.97±0.04	4.84±0.04	-0.138	-2.77
liquid line temperature drop (°C)	ΔT_{LL}	2.18±0.05	2.27±0.05	0.090	4.11
outdoor temperature minus TC#103 (°C)	ΔT_{103}	6.98±0.05	7.07±0.03	0.088	1.26
ID fan motor case temperature (°C)	T_{IDF}	67.02±0.12	66.28±0.03	-0.737	-1.10
OD fan motor case temperature (°C)	T_{ODF}	30.13±0.04	30.14±0.02	0.007	0.02
reversing valve temperature change, discharge side (°C)	ΔT_{RVD}	3.26±0.14	3.39±0.14	0.135	4.13
reversing valve temperature change, suction side (°C)	ΔT_{RVS}	0.75±0.08	0.74±0.04	-0.012	-1.65

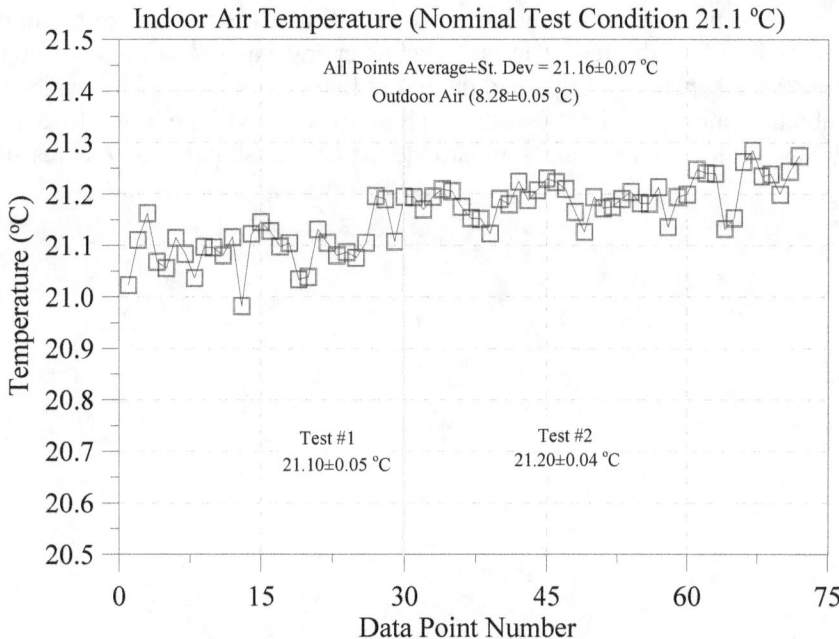

Figure 4.5.5. Variation of indoor air dry-bulb temperature for several test cases with nominal conditions of 21.1 °C indoor and 8.3 °C outdoor dry-bulb temperatures

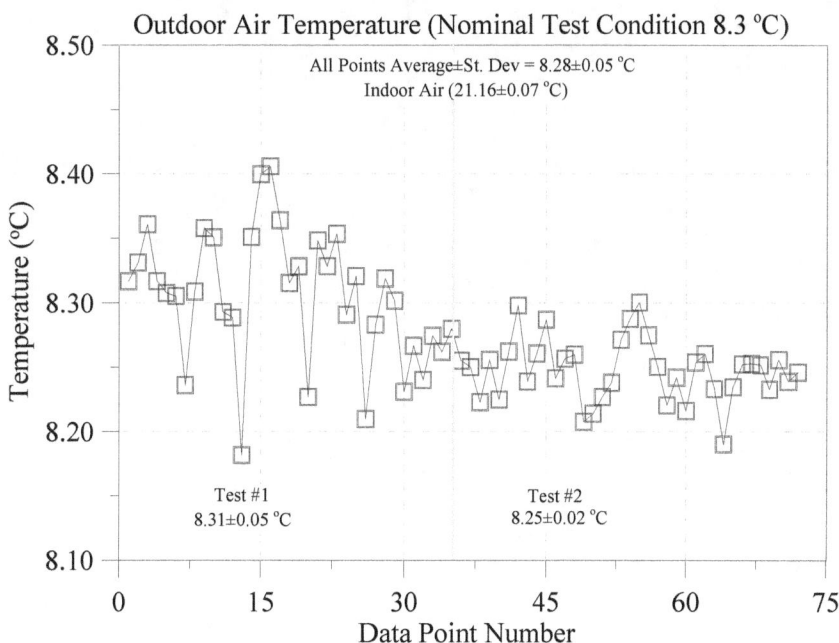

Figure 4.5.6. Variation of outdoor air dry-bulb temperature for several test cases with nominal conditions of 21.1 °C indoor and 8.3 °C outdoor dry-bulb temperatures

Table 4.5.4 compares selected features and repeat tests (Tests #1, #2 and #3) for a given nominal test condition of 21.1 °C indoor and 17.6 °C outdoor dry-bulb temperatures. The highest percent difference between the reference test and the repeat feature values occurred for ΔT_{RVS}, ΔT_{LL}, ΔT_{scV}, T_{ER}, and ΔT_{shE}. Figures 4.5.7 and 4.5.8 graphically show the test temperature conditions.

Slight variations in ambient conditions yielded more significant variation in refrigerant vapor line and liquid line conditions in all of the repeat tests. The reversing valve suction side temperature change, ΔT_{RVS}, the evaporator exit superheat, ΔT_{shE}, and the liquid line subcooling, ΔT_{scV}, tended to show the most sensitivity to ambient conditions for all tests. These results will be used to set the steady-state temperature variation limits (steady-state standard deviation thresholds) for the real-time steady-state detector.

Table 4.5.4. Variation in features due to variation in ambient condition repeatability for a given nominal test condition of 21.1 °C indoor and 16.7 °C outdoor dry-bulb temperatures

Feature Name	Feature Symbol	Ref. Value	Repeat Test #2 Value	Residual	% Diff. wrt the Reference	Repeat Test #3 Value	Residual	% Diff. wrt the Reference
total air side capacity (kW)	Q_{CA}	9.754±0.038	9.702±0.027	-0.052	-0.54	9.705±0.045	-0.049	-0.50
compressor power (kW)	W_{comp}	1.917±0.003	1.913±0.002	-0.003	-0.17	1.913±0.003	-0.004	-0.20
refrigerant mass flow rate (kg min^{-1})	m_R	2.965±0.012	2.939±0.013	-0.026	-0.88	2.938±0.024	-0.027	-0.90
coefficient of performance	COP	3.934±0.021	3.931±0.019	-0.003	-0.07	3.929±0.032	-0.005	-0.12
indoor unit refrigerant side capacity (kW)	Q_{CR}	9.705±0.045	9.690±0.047	-0.015	-0.16	9.686±0.082	-0.019	-0.19
evaporator exit saturation temperature (°C)	T_{ER}	6.46±0.01	6.17±0.05	-0.297	-4.60	6.14±0.16	-0.325	-5.03
evaporator bend thermocouple, TC#103 (°C)	T_{E103}	8.36±0.02	8.11±0.03	-0.249	-2.98	8.07±0.17	-0.294	-3.52
evaporator exit superheat (°C)	ΔT_{shE}	7.87±0.05	8.29±0.05	0.422	5.36	8.22±0.15	0.355	4.51
compressor discharge wall temperature (°C)	T_D	63.42±0.10	63.78±0.08	0.362	0.57	63.76±0.21	0.344	0.54
condenser inlet saturation temperature (°C)	T_{CR}	41.72±0.07	41.48±0.04	-0.238	-0.57	41.46±0.09	-0.265	-0.64
condenser bend thermocouple, TC#15 (°C)	T_{C15}	42.49±0.09	42.42±0.06	-0.067	-0.16	42.41±0.11	-0.080	-0.19
condenser inlet superheat (°C)	ΔT_{shC}	23.80±0.19	24.32±0.19	0.524	2.20	24.26±0.26	0.460	1.93
vapor superheat at outdoor service valve (°C)	ΔT_{shV}	25.47±0.26	25.95±0.24	0.477	1.87	25.94±0.29	0.468	1.84
liquid line subcooling at outdoor service valve (°C)	ΔT_{scV}	2.49±0.07	2.77±0.05	0.277	11.12	2.77±0.09	0.278	11.15
condenser air temperature rise (°C)	ΔT_{CA}	17.55±0.06	17.69±0.04	0.140	0.80	17.70±0.07	0.150	0.85
evaporator air temperature drop (°C)	ΔT_{EA}	6.33±0.04	6.17±0.04	-0.160	-2.53	6.13±0.05	-0.196	-3.10
liquid line temperature drop (°C)	ΔT_{LL}	1.91±0.04	1.81±0.04	-0.106	-5.53	1.81±0.04	-0.101	-5.28
outdoor temperature minus TC#103 (°C)	ΔT_{103}	8.35±0.04	8.57±0.05	0.212	2.53	8.52±0.12	0.168	2.02
ID fan motor case temperature (°C)	T_{IDF}	69.69±0.19	70.08±0.03	0.384	0.55	70.07±0.10	0.372	0.53
OD fan motor case temperature (°C)	T_{ODF}	37.63±0.05	37.11±0.11	-0.522	-1.39	37.34±0.07	-0.286	-0.76
reversing valve temperature change, discharge side (°C)	ΔT_{RVD}	3.62±0.24	3.54±0.25	-0.076	-2.12	3.52±0.23	-0.096	-2.66
reversing valve temperature change, suction side (°C)	ΔT_{RVS}	0.24±0.07	0.28±0.05	0.035	14.57	0.31±0.16	0.066	27.35

67

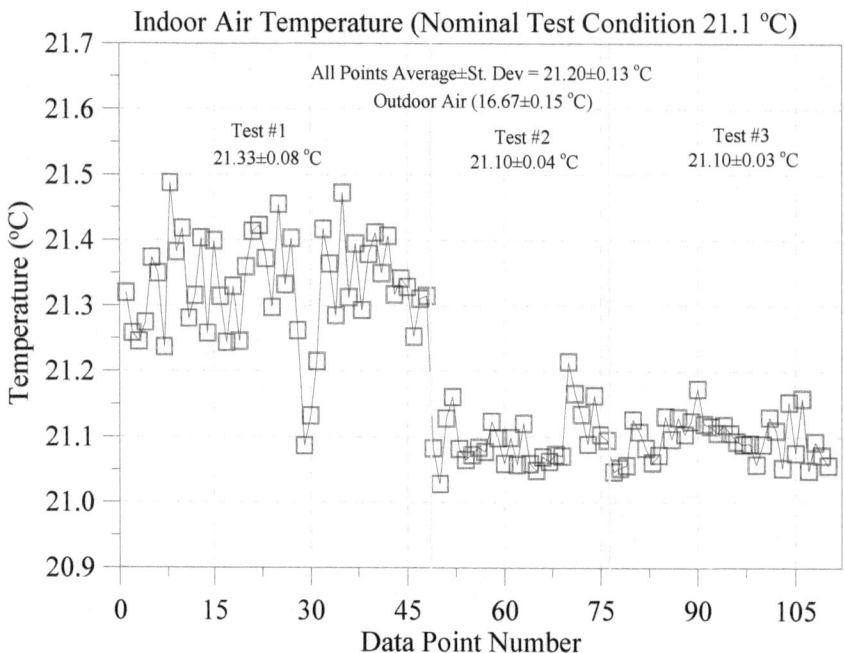

Figure 4.5.7. Variation of indoor air dry-bulb temperature for several test cases with nominal conditions of 21.1 °C indoor and 16.7 °C outdoor dry-bulb temperatures

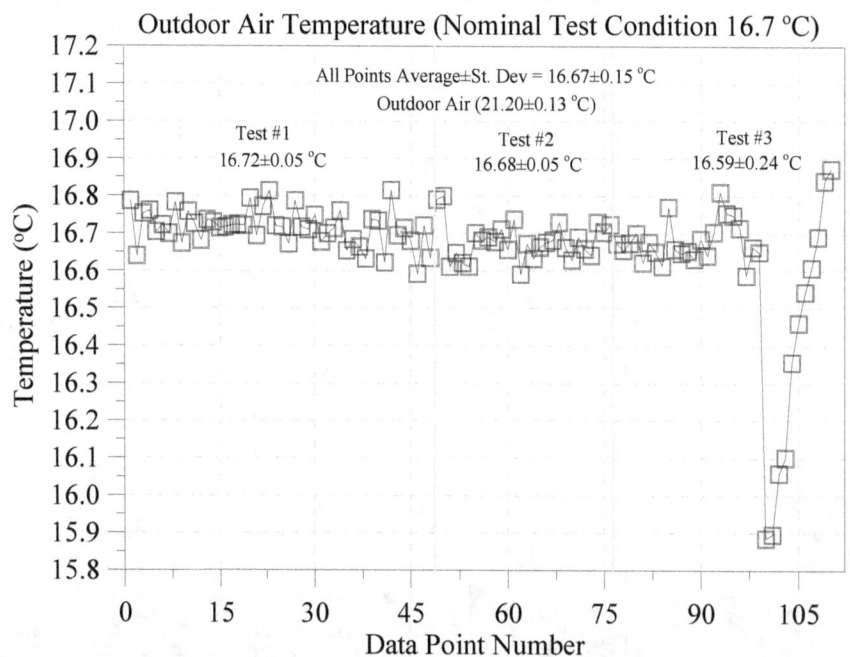

Figure 4.5.8. Variation of outdoor air dry-bulb temperature for several test cases with nominal conditions of 21.1 °C indoor and 16.7 °C outdoor dry-bulb temperatures

CHAPTER 5. Single-Fault Test Results

A constant indoor dry-bulb temperature of 21.1 °C at outdoor conditions of -8.3 °C/Dry and 8.3 °C/(72.5 % RH and Dry) were set while the following six faults were imposed: 1) condenser or indoor coil improper airflow, 2) evaporator or outdoor coil improper airflow, 3) compressor or four-way valve leakage, 4) refrigerant liquid line restriction, 5) refrigerant overcharge, and 6) refrigerant undercharge. The fault level is defined by the percent change from the reference case. Residuals were calculated according to Equation 4.5.1 as the fault imposed value minus the fault-free value at the same test condition.

5.1 Condenser Air Flow Fault (CF fault)

5.1.1   Indoor dry-bulb temperature of 21.1 °C at outdoor conditions of -8.3 °C/Dry

The air flow rate through the indoor coil was varied by changing the speed of the nozzle chamber booster fan. Figure 5.1.1 shows the change in heating capacity, refrigerant-side heating capacity, compressor power, COP, and refrigerant mass flow rate as a function of the percent reduction in indoor air flow rate. COP was calculated based upon refrigerant-side heating capacity. Disagreement between refrigerant and air side heating capacities increased as air flow rate was reduced, but their agreement remained better than 6 %.

Figure 5.1.2 shows the residuals of T_{ER}, T_{E103}, and ΔT_{shE} as a function of indoor coil air flow rate percent reduction. The sudden downward jump in evaporator exit superheat was caused by a change in mass flow rate induced by the actions of the TXV; an increase in refrigerant mass flow rate caused the decrease in superheat. Increasing the fault level caused the evaporator exit superheat to begin increasing at the 30 % fault level. The non-linear sawtooth pattern of all the residuals was caused by the TXV moving to correct evaporator exit superheat.

Figure 5.1.3 shows residuals for T_D, T_{CR}, T_{C15}, T_{shC}, T_{shV}, and T_{scV}. All residuals show positive slopes as fault level increased. The largest slope occured with the compressor discharge wall temperature; an absolute change of more than 2.75 °C occurs at the highest fault level. Note that the condenser inlet saturation temperature, calculated from a pressure measurement, was closely paralleled by the temperature measured by thermocouple, T_{C15}. Superheat residuals were again fluctuating due to corrections applied to refrigerant mass flow rate by the TXV.

Figure 5.1.4 shows residuals for ΔT_{CA}, ΔT_{EA}, ΔT_{LL}, and ΔT_{103}. As expected, the indoor coil air temperature change increased with decreases in the air flow rate; the maximum change was 3.0 °C at the highest fault level. The remaining parameters showed less than 0.5 °C change at the highest fault level.

Figure 5.1.5 shows residuals for T_{IDF}, T_{ODF}, ΔT_{RVD}, and ΔT_{RVS}. As indoor air flow rate decreased the indoor fan's case temperature increased; a maximum change of almost 8.0 °C occurred at the highest fault level. Temperature change around the reversing valve and the outdoor fan case was relatively negligible; less than 0.5 °C.

Table 5.1.1 shows the linear slopes and absolute value of the percent changes in the system characteristics and residual temperatures for the indoor air flow faults. Seven features changed by more than 5 % at the maximum fault level; ΔT_{CA}, T_{IDF}, ΔT_{LL}, ΔT_{RVD}, T_{ER}, T_{C15}, and T_{CR}. The residuals which had the largest slopes are important to indicating the condenser air flow. The five largest changes in the residuals with increased indoor coil air flow reduction occurred for T_{IDF}, T_D, ΔT_{CA}, T_{C15}, and T_{CR}.

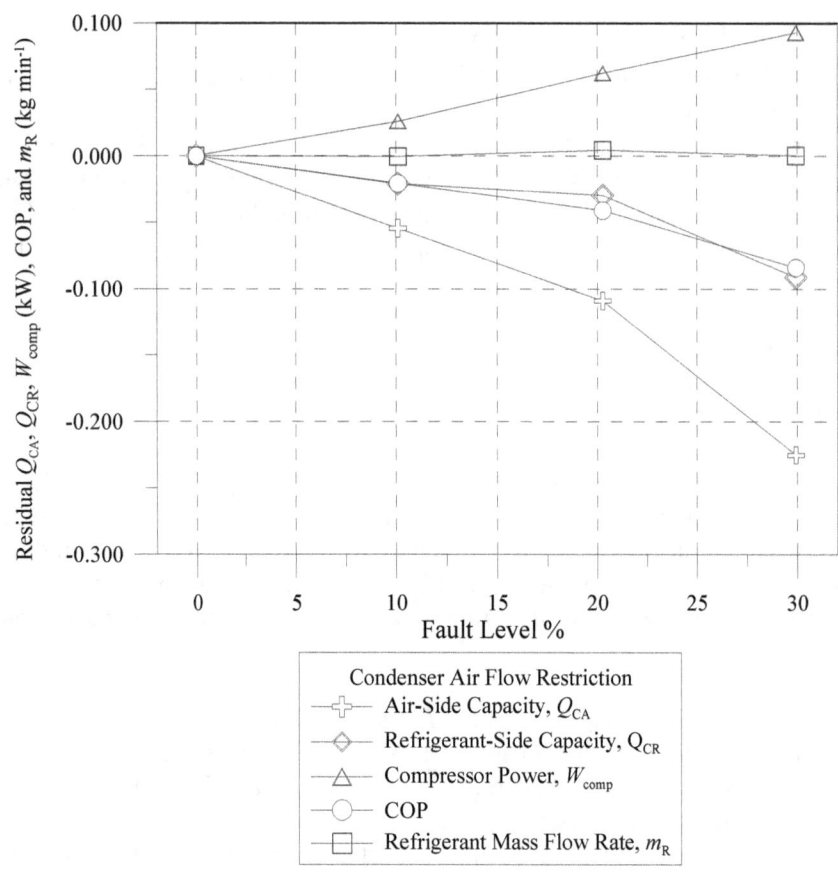

Figure 5.1.1. Residual of selected features with a condenser fouling fault at an indoor dry-bulb temperature of 21.1 °C and outdoor conditions of -8.3 °C/Dry: R[Q_{CA}], R[Q_{CR}], R[W_{comp}], R[COP], and R[m_R]

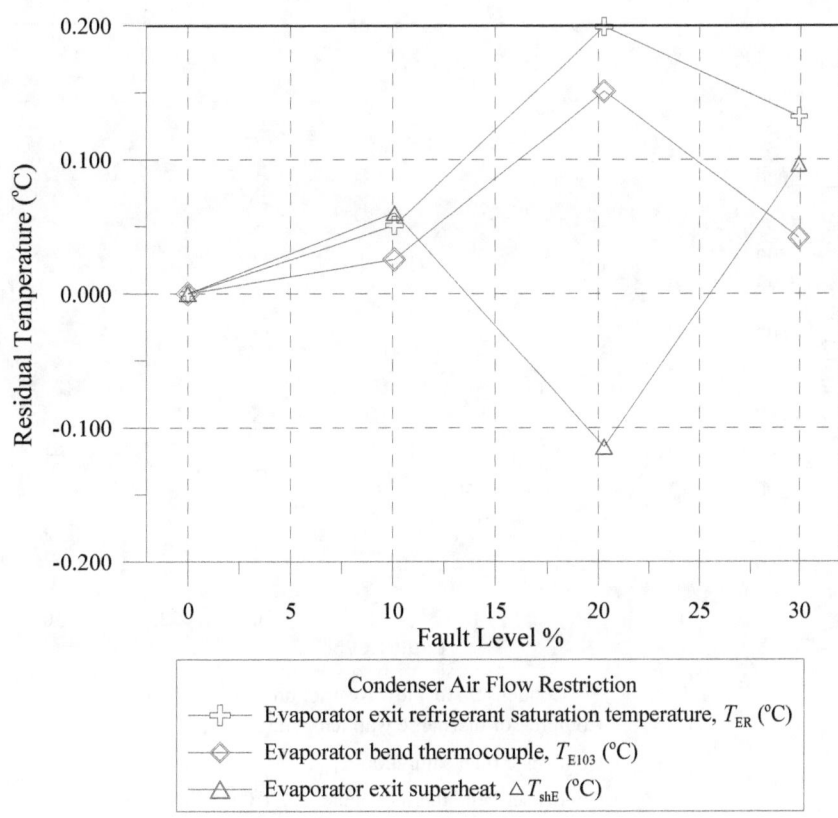

Figure 5.1.2. Residual of selected features with a condenser fouling fault at an indoor dry-bulb temperature of 21.1 °C and outdoor conditions of -8.3 °C/Dry: R[T_{ER}], R[T_{E103}], and R[ΔT_{shE}]

71

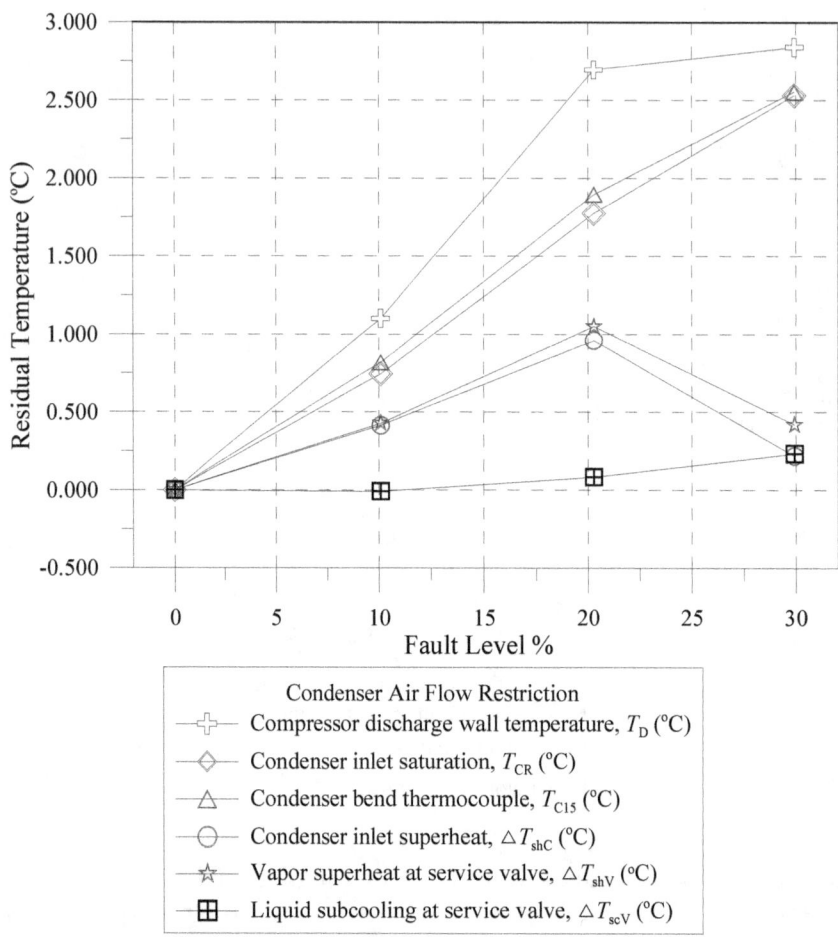

Figure 5.1.3. Residual of selected features with a condenser fouling fault at an indoor dry-bulb
temperature of 21.1 °C and outdoor conditions of -8.3 °C/Dry: R[T_D], R[T_{CR}], R[T_{C15}],
R[ΔT_{shC}], R[ΔT_{shV}], and R[ΔT_{scV}]

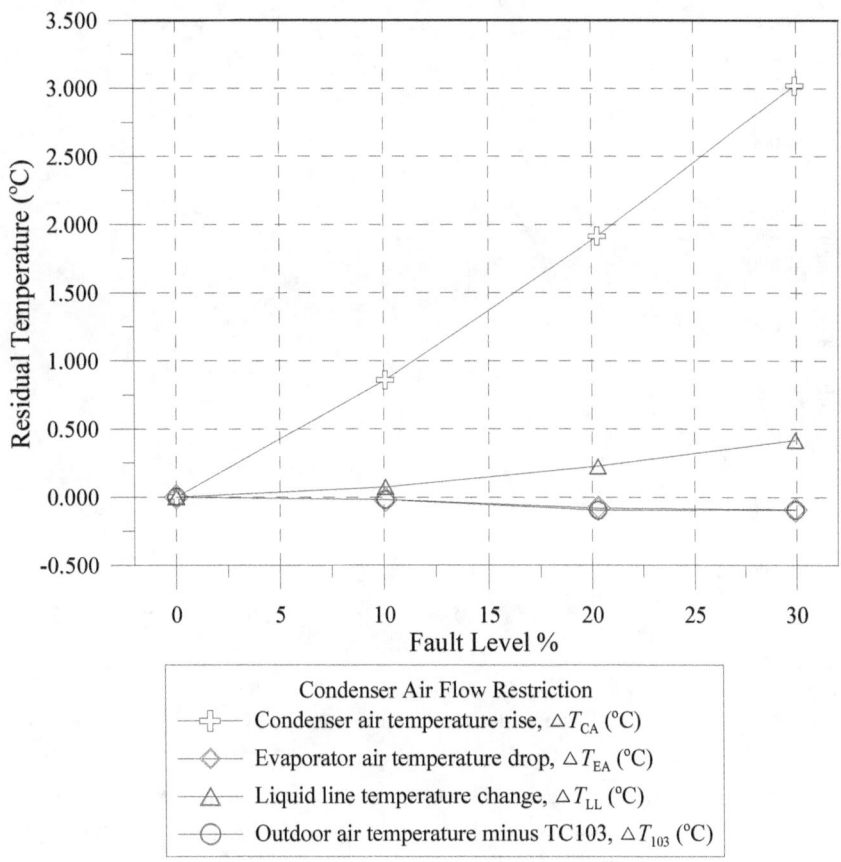

Figure 5.1.4. Residual of selected features with a condenser fouling fault at an indoor dry-bulb temperature of 21.1 °C and outdoor conditions of -8.3 °C/Dry: R[ΔT_{CA}], R[ΔT_{EA}], R[ΔT_{LL}], and R[ΔT_{103}]

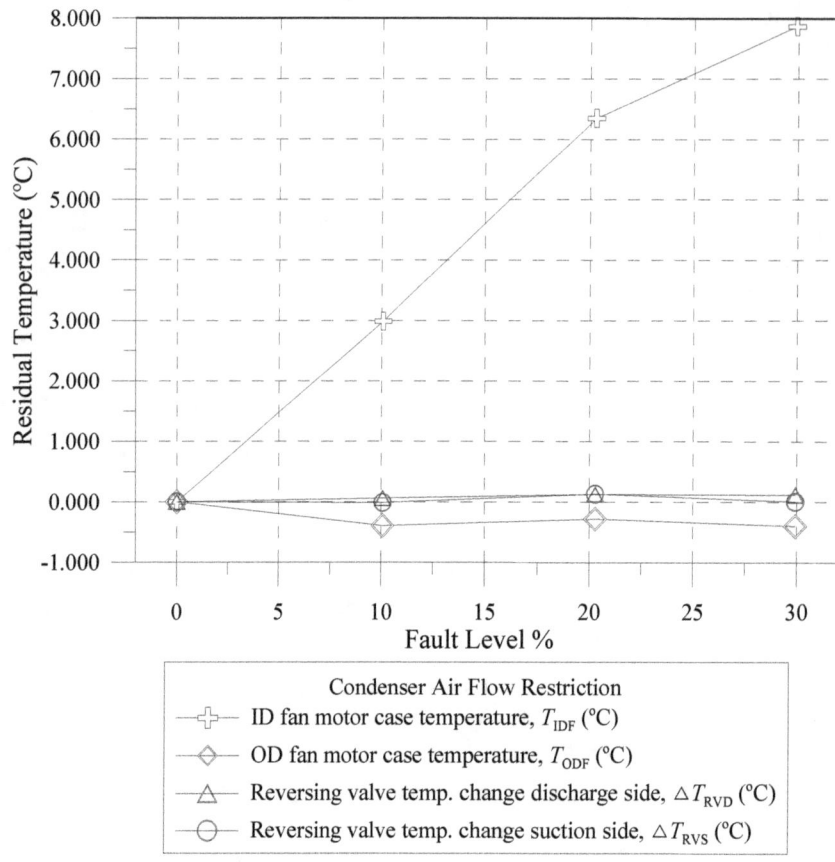

Figure 5.1.5. Residual of selected features with a condenser fouling fault at an indoor dry-bulb temperature of 21.1 °C and outdoor conditions of -8.3 °C/Dry: R[T_{IDF}], R[T_{ODF}], R[ΔT_{RVD}], and R[ΔT_{RVS}]

Table 5.1.1. Linear fit slopes of features as a function of percent condenser air flow rate fault at an indoor dry-bulb temperature of 21.1 °C and outdoor conditions of -8.3 °C/Dry

Feature Name	Feature Symbol	Feature's slope	Feature % Change @ Max Fault Level
total air side capacity (kW %$^{-1}$)	Q_{CA}	-7.28E-03	-4.4
compressor power (kW %$^{-1}$)	W_{comp}	3.15E-03	5.8
refrigerant mass flow rate (kg min^{-1} %$^{-1}$)	m_R	6.07E-05	0.0
coefficient of performance(%-1)	COP	-2.72E-03	-3.7
indoor unit refrigerant side capacity (kW %$^{-1}$)	Q_{CR}	-2.81E-03	-1.8

Fault imposed by lowering nozzle chamber booster fan drive frequency. Fault level equal to % decrease in indoor coil air flow rate.

Max Fault Level: 29.94 %

Listed in Descending Order of Largest ABS(Δ°C (%Fault)$^{-1}$)

		Residual's slope as a function of % fault level	
		Δ°C (%Fault)$^{-1}$	Feature % Change @ Max Fault Level
ID fan motor case temperature (°C)	T_{IDF}	0.270	10.1
compressor discharge wall temperature (°C)	T_D	0.102	3.7
condenser air temperature rise (°C)	ΔT_{CA}	0.101	32.4
condenser bend thermocouple, TC#15 (°C)	T_{C15}	0.087	5.1
condenser inlet saturation temperature (°C)	T_{CR}	0.086	5.1
vapor superheat at outdoor service valve (°C)	ΔT_{shV}	0.019	1.4
liquid line temperature drop (°C)	ΔT_{LL}	0.014	9.9
condenser inlet superheat (°C)	ΔT_{shC}	0.012	0.9
OD fan motor case temperature (°C)	T_{ODF}	-0.011	-1.1
liquid line subcooling at outdoor service valve (°C)	ΔT_{scV}	0.008	3.6
evaporator exit saturation temperature (°C)	T_{ER}	0.005	5.7
reversing valve temperature change, discharge side (°C)	ΔT_{RVD}	0.004	6.6
outdoor temperature minus TC#103 (°C)	ΔT_{103}	-0.004	-1.9
evaporator air temperature drop (°C)	ΔT_{EA}	-0.003	-3.3
evaporator bend thermocouple, TC#103 (°C)	T_{E103}	0.003	0.9
reversing valve temperature change, suction side (°C)	ΔT_{RVS}	0.002	0.2
evaporator exit superheat (°C)	ΔT_{shE}	0.001	1.6

5.1.2 Indoor Dry-Bulb Temperature of 21.1 °C at Outdoor Conditions of 8.3 °C/72.5 % RH

Figure 5.1.6 shows the change in air-side heating capacity, refrigerant-side heating capacity, compressor power, COP, and refrigerant mass flow rate as a function of the percent reduction in indoor air flow rate. COP decreased by almost 10 % with a 31 % reduction in indoor air flow rate. Air-side and refrigerant-side capacity decreased by more than 8 % and 4 %, respectively.

Figure 5.1.7 shows the residuals of T_{ER}, T_{E103}, and ΔT_{shE} as a function of indoor coil air flow rate percent reduction. Evaporator exit superheat residual showed a greater change than the evaporator exit saturation temperature residual, but both features changed with decreasing indoor air flow rate. The changes in residual slopes were not significant relative to other feature's residual slopes.

Figure 5.1.8 shows residuals for T_D, T_{CR}, T_{C15}, ΔT_{shC}, ΔT_{shV}, and ΔT_{scV}. All residuals, except for the outdoor service valve liquid line, showed positive slopes with decreasing indoor coil air flow rate. The largest slope occured for the residual of the compressor discharge wall temperature; an absolute change of more than 6.0 °C occurs at the highest fault level. Note that the condenser inlet saturation temperature, calculated from a pressure measurement, was closely paralleled by the temperature measured by thermocouple, T_{C15}. Superheats and subcooling were also affected by the reduced indoor coil air flow.

Figure 5.1.9 shows residuals for ΔT_{CA}, ΔT_{EA}, ΔT_{LL}, and ΔT_{103}. As expected, the indoor coil air temperature change increased with decreases in the air flow rate; the maximum change was 4.2 °C at the highest fault level. The other features showed less than 0.5 °C change at the highest fault level.

Figure 5.1.10 shows residuals for T_{IDF}, T_{ODF}, ΔT_{RVD}, and ΔT_{RVS}. As indoor air flow rate decreased the indoor fan's case temperature increased; a maximum change of almost 11.0 °C occurred at the highest fault level. Temperature change around the reversing valve and the outdoor fan case was negligible; less than 0.6 °C.

Table 5.1.2 shows the linear slopes and absolute value of the percent changes in the system features and residual temperatures for indoor air flow faults. Five features showed residual changes with fault level of more than 0.1 °C %$^{-1}$; T_{IDF}, T_D, T_{C15}, ΔT_{CA}, and T_{CR}. Of these features the air temperature change across the condenser (indoor coil) had the greatest change at the maximum fault level, 30.7 %. The residuals with the largest slope were indoor fan case temperature and the compressor discharge line wall temperature with values of 0.351 °C %$^{-1}$ and 0.209 °C %$^{-1}$.

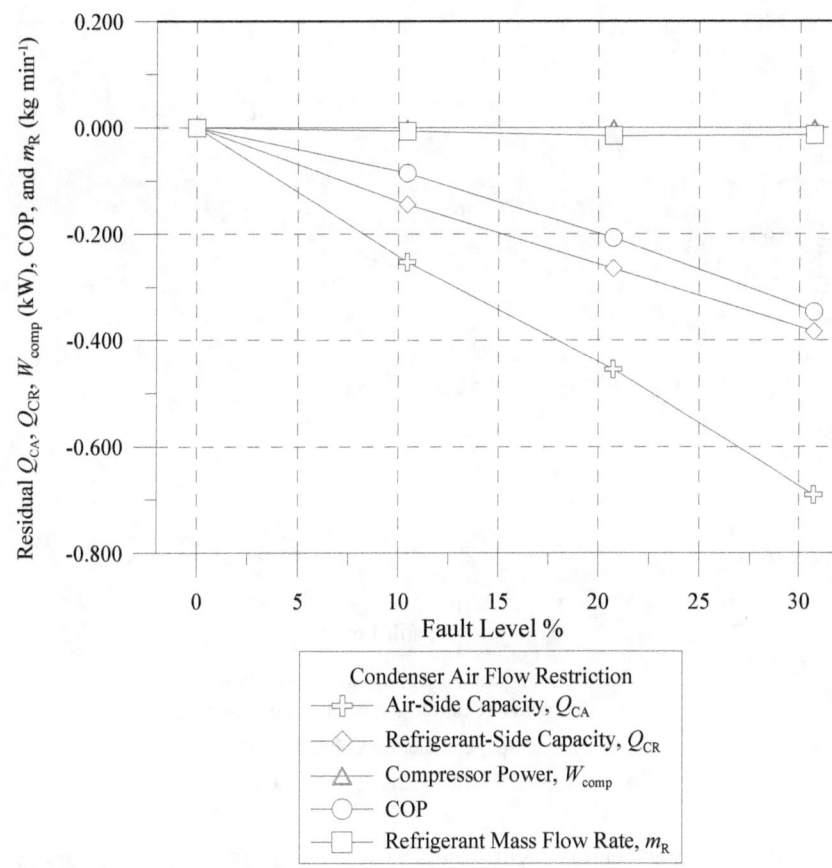

Figure 5.1.6. Residual of selected features with a condenser fouling fault at an indoor dry-bulb temperature of 21.1 °C and outdoor conditions of 8.3 °C/72.5 % RH: R[Q_{CA}], R[Q_{CR}], R[W_{comp}], R[COP], and R[m_R]

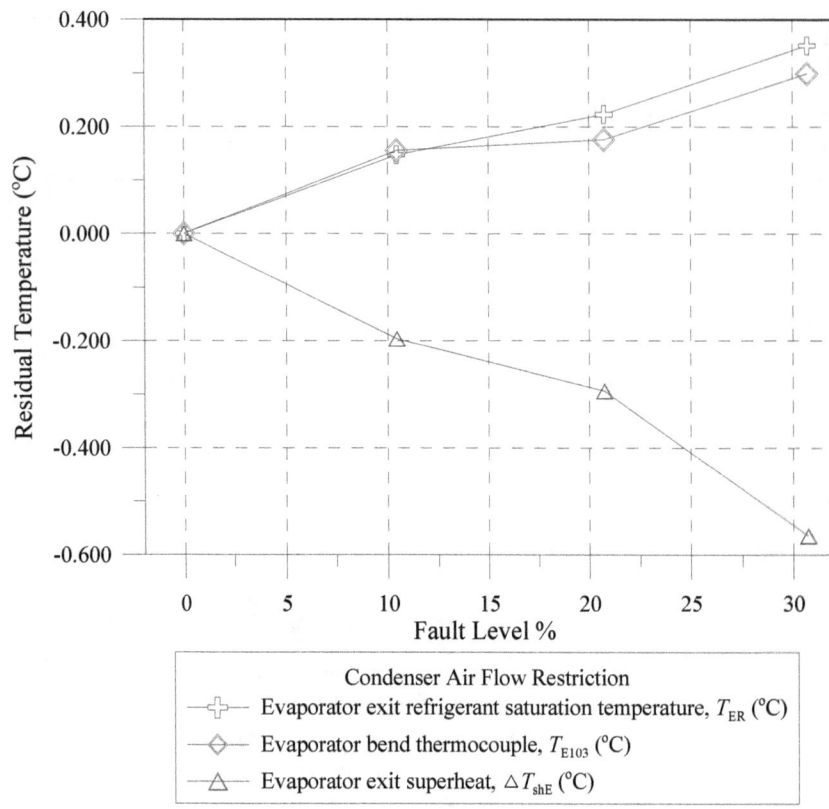

Figure 5.1.7. Residual of selected features with a condenser fouling fault at an indoor dry-bulb temperature of 21.1 °C and outdoor conditions of 8.3 °C/72.5 % RH: R[T_{ER}], R[T_{E103}], and R[ΔT_{shE}]

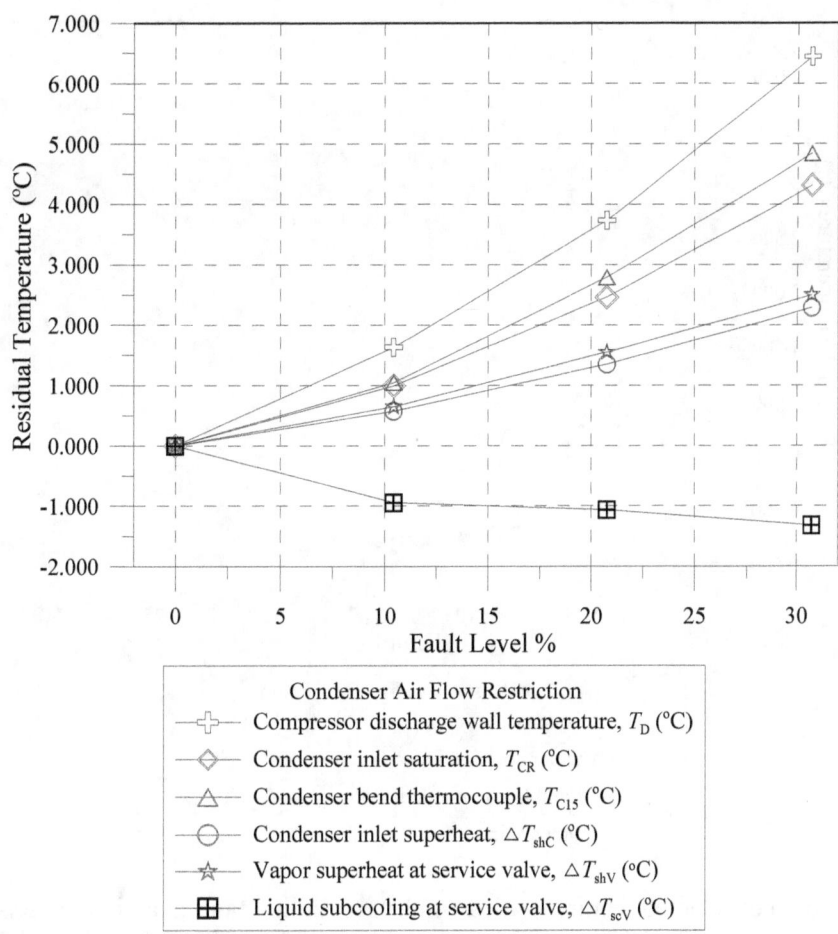

Figure 5.1.8. Residual of selected features with a condenser fouling fault at an indoor dry-bulb temperature of 21.1 °C and outdoor conditions of 8.3 °C/72.5 % RH: R[T_D], R[T_{CR}], R[T_{C15}], R[ΔT_{shC}], R[ΔT_{shV}], and R[ΔT_{scV}]

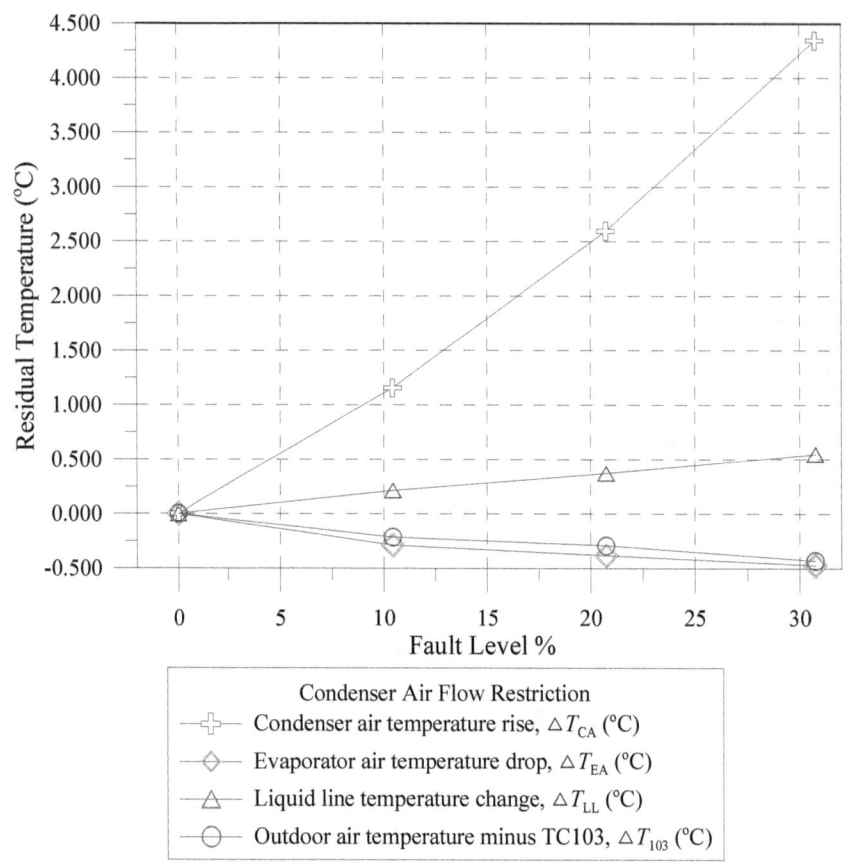

Figure 5.1.9. Residual of selected features with a condenser fouling fault at an indoor dry-bulb temperature of 21.1 °C and outdoor conditions of 8.3 °C/72.5 % RH: $R[\Delta T_{CA}]$, $R[\Delta T_{EA}]$, $R[\Delta T_{LL}]$, and $R[\Delta T_{103}]$

80

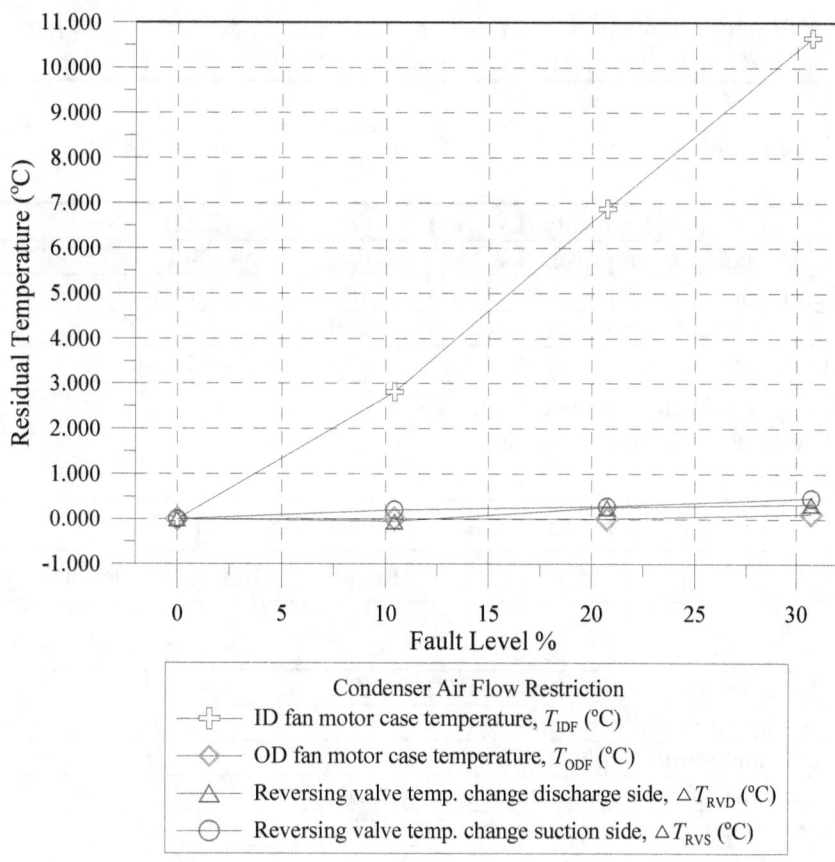

Figure 5.1.10. Residual of selected features with a condenser fouling fault: R[T_{IDF}], R[T_{ODF}], R[ΔT_{RVD}], and R[ΔT_{RVS}] at an indoor dry-bulb temperature of 21.1 °C and outdoor conditions of 8.3 °C/72.5 % RH

Table 5.1.2. Linear fit slopes of features as a function of percent condenser air flow rate fault for a given nominal test conditions of 21.1 °C indoor and 8.3 °C/72.5 % RH outdoor

Feature Name	Feature Symbol	Feature's slope	Feature % Change @ Max Fault Level
total air side capacity (kW %$^{-1}$)	Q_{CA}	-2.22E-02	-8.3
compressor power (kW %$^{-1}$)	W_{comp}	6.40E-03	10.8
refrigerant mass flow rate (kg min^{-1} %$^{-1}$)	m_R	-4.95E-04	-0.6
coefficient of performance(%-1)	COP	-1.13E-02	-10.0
indoor unit refrigerant side capacity (kW %$^{-1}$)	Q_{CR}	-1.24E-02	-4.6

Fault imposed by lowering nozzle chamber booster fan drive frequency. Fault level equal to % decrease in indoor coil standard air flow rate.		Max Fault Level: 30.7 %

Listed in Descending Order of Largest ABS(Δ°C (%Fault)$^{-1}$)

		Residual's slope as a function of % fault level	
		Δ°C (%Fault)$^{-1}$	Feature % Change @ Max Fault Level
ID fan motor case temperature (°C)	T_{IDF}	0.351	12.6
compressor discharge wall temperature (°C)	T_D	0.209	8.2
condenser bend thermocouple, TC#15 (°C)	T_{C15}	0.158	8.5
condenser air temperature rise (°C)	ΔT_{CA}	0.141	28.9
condenser inlet saturation temperature (°C)	T_{CR}	0.140	7.7
vapor superheat at outdoor service valve (°C)	ΔT_{shV}	0.082	9.9
condenser inlet superheat (°C)	ΔT_{shC}	0.074	9.9
liquid line subcooling at outdoor service valve (°C)	ΔT_{scV}	-0.040	-32.7
liquid line temperature drop (°C)	ΔT_{LL}	0.018	24.5
evaporator exit superheat (°C)	ΔT_{shE}	-0.017	-9.2
evaporator air temperature drop (°C)	ΔT_{EA}	-0.015	-11.7
reversing valve temperature change, suction side (°C)	ΔT_{RVS}	0.014	42.2
outdoor temperature minus TC#103 (°C)	ΔT_{103}	-0.013	-6.8
reversing valve temperature change, discharge side (°C)	ΔT_{RVD}	0.013	10.3
evaporator exit saturation temperature (°C)	T_{ER}	0.011	1.9
evaporator bend thermocouple, TC#103 (°C)	T_{E103}	0.009	1.5
OD fan motor case temperature (°C)	T_{ODF}	0.003	0.2

5.2 Evaporator Air Flow Fault (EF fault)

5.2.1   Indoor dry-bulb temperature of 21.1 °C at outdoor conditions of -8.3 °C/Dry

The air flow rate through the outdoor coil was varied by placing solid ribbons of paper across the bottom edge face of the coil, thus increasing the flow resistance, and blocking air flow along the bottom circuits (also see Figure 4.2.3.6). The bottom circuits are most likely to be blocked due to snow, icing, or dirt/trash accumulation. Figure 5.2.1 shows the change in air-side heating capacity, refrigerant-side heating capacity, compressor power, COP, and refrigerant mass flow rate as a function of the percent blockage in outdoor coil face area. Refrigerant-side heating capacity and COP decreased by 26.5 % and

23.5 %, respectively, with a 30 % blockage in outdoor coil face area causing a 14.3 % increase in coil pressure drop.

Figure 5.2.2 shows the residuals of T_{ER}, T_{E103}, and ΔT_{shE} as a function of outdoor coil area blockage. At the maximum fault level, the evaporator exit refrigerant saturation temperature, T_{ER}, and the evaporator exit superheat, ΔT_{shE}, changed by more than 100 % from their NFSS values; the residuals of these two features changed more than any other feature with slope values of -0.261 C %$^{-1}$ and 0.250 C %$^{-1}$, respectively.

Figure 5.2.3 shows residuals for T_D, T_{CR}, T_{C15}, ΔT_{shC}, ΔT_{shV}, and ΔT_{scV}. Compressor discharge line wall temperature, T_D, and condenser inlet saturation temperature, T_{CR}, residuals show the largest change; increasing and decreasing by more than 4 % and 6 %, respectively, from their NFSS values.

Figure 5.2.4 shows residuals for ΔT_{CA}, ΔT_{EA}, ΔT_{LL}, and ΔT_{103}. At the maximum fault level, ΔT_{103} residual increases by more than 5.5 °C, which corresponds to a 116 % increase of this feature over its NFSS value. The condenser air temperature rise residual also showed a noticeable change resulting from a 23 % decrease from it NFSS value.

Figure 5.2.5 shows residuals for T_{IDF}, T_{ODF}, ΔT_{RVD}, and ΔT_{RVS}. As outdoor coil area blockage increased the indoor fan's case temperature decreased more than the outdoor fan's case temperature. The maximum change occurred for the residual of ΔT_{RVS}; decreasing by more than 7 °C, which corresponded to a 350 % decrease in this feature from its NFSS value.

Table 5.2.1 shows the residual's linear slopes and the percent changes in the system characteristics and temperatures for the evaporator or outdoor air flow fault. Thirteen features changed by more than 5 % at the maximum fault level; the top five were: ΔT_{RVS}, T_{ER}, ΔT_{RVD}, ΔT_{shE}, and ΔT_{103} (in descending order of maximum magnitude change). The residuals that have the largest changes as a function of fault level are important to indicating an outdoor coil air flow fault; T_{ER}, ΔT_{shE}, ΔT_{RVS}, T_{E103}, and ΔT_{103} had residual slopes greater than 0.18 C %$^{-1}$.

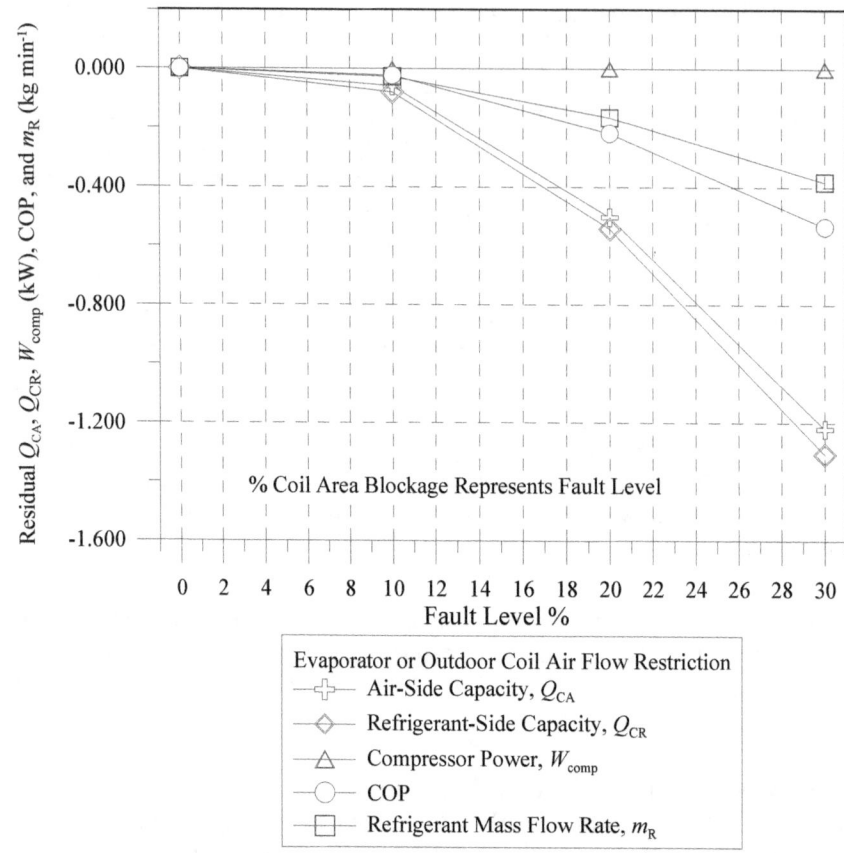

Figure 5.2.1. Residual of selected features with evaporator air flow faults at an indoor dry-bulb temperature of 21.1 °C and outdoor conditions of -8.3 °C/Dry: R[Q_{CA}], R[Q_{CR}], R[W_{comp}], R[COP], and R[m_R]

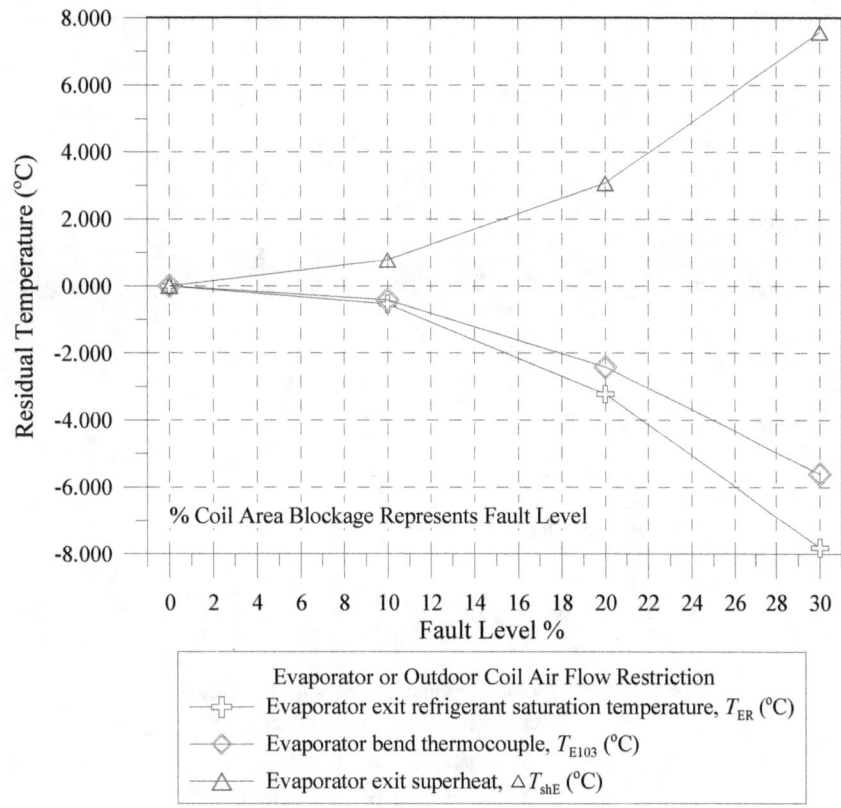

Figure 5.2.2. Residual of selected features with an evaporator fouling fault at an indoor dry-bulb temperature of 21.1 °C and outdoor conditions of -8.3 °C/Dry: R[T_{ER}], R[T_{E103}], and R[ΔT_{shE}]

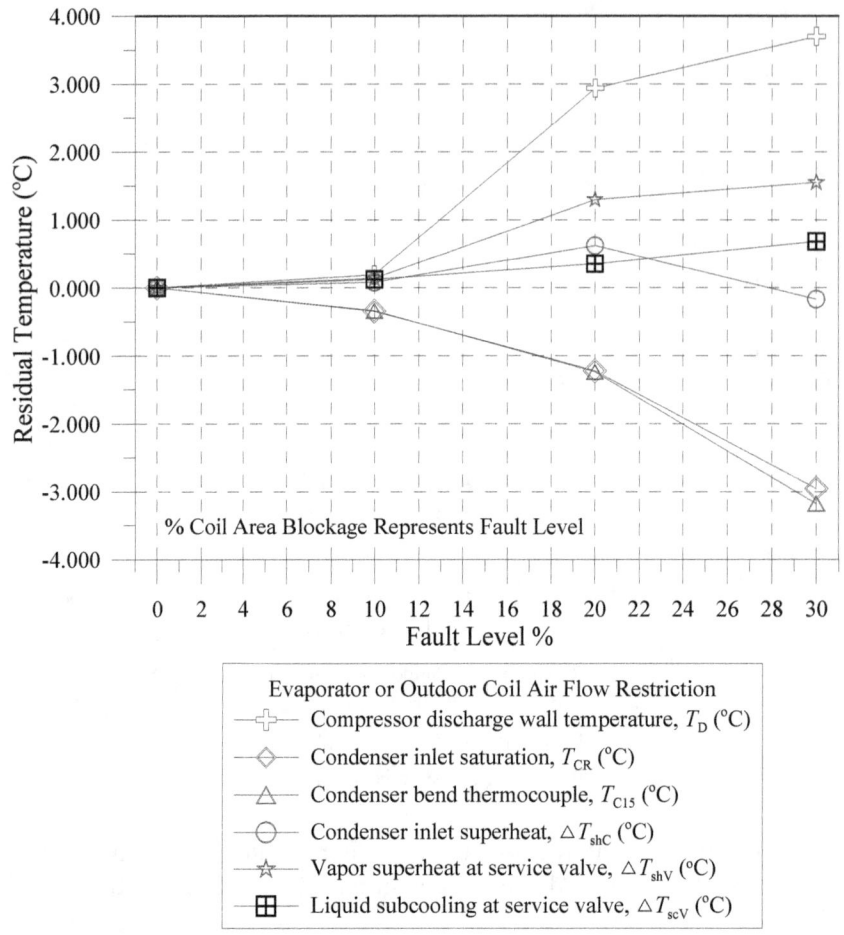

Figure 5.2.3. Residual of selected features with an evaporator fouling fault at an indoor dry-bulb temperature of 21.1 °C and outdoor conditions of -8.3 °C/Dry: R[T_D], R[T_{CR}], R[T_{C15}], R[ΔT_{shC}], R[ΔT_{shV}], and R[ΔT_{scV}]

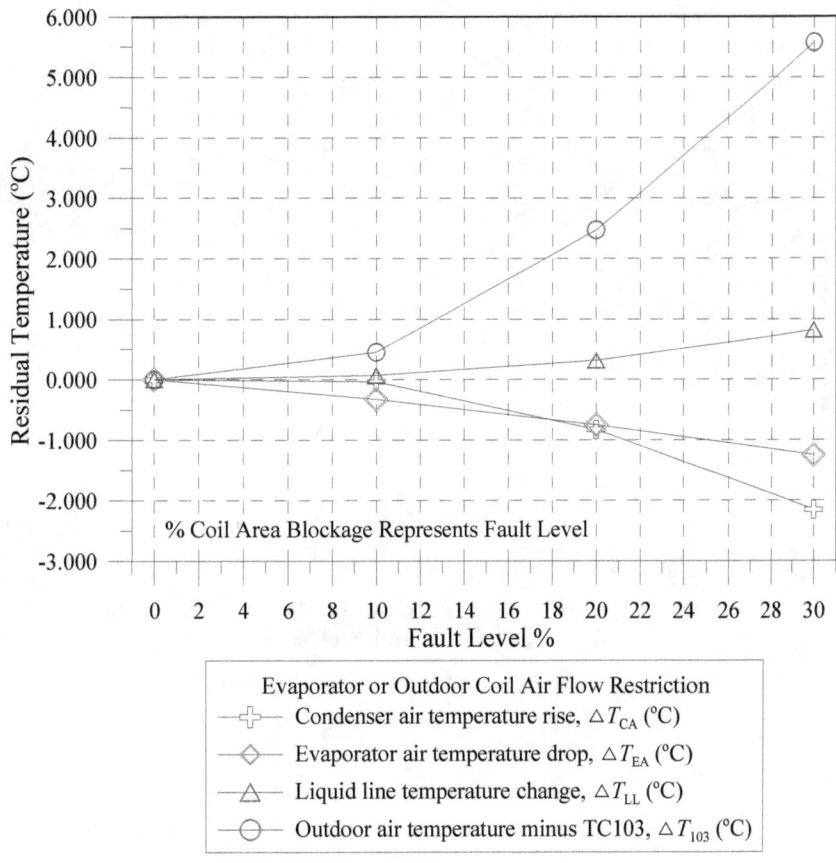

Figure 5.2.4. Residual of selected features with an evaporator fouling fault at an indoor dry-bulb temperature of 21.1 °C and outdoor conditions of -8.3 °C/Dry: $R[\Delta T_{CA}]$, $R[\Delta T_{EA}]$, $R[\Delta T_{LL}]$, and $R[\Delta T_{103}]$

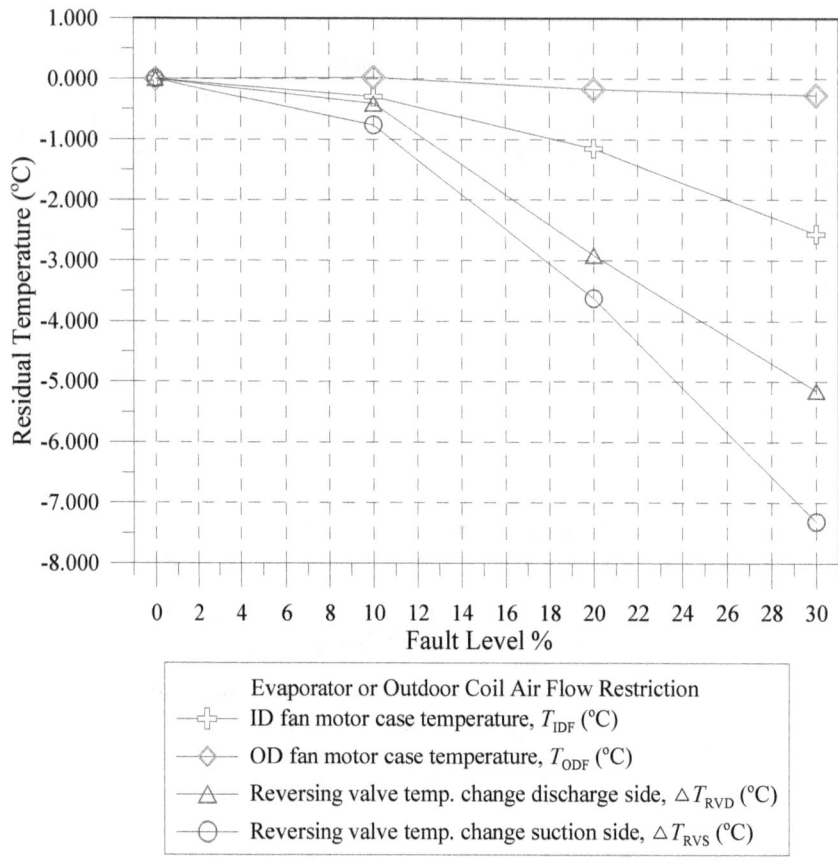

Figure 5.2.5. Residual of selected features with an evaporator fouling fault at an indoor dry-bulb temperature of 21.1 °C and outdoor conditions of -8.3 °C/Dry: $R[T_{IDF}]$, $R[T_{ODF}]$, $R[\Delta T_{RVD}]$, and $R[\Delta T_{RVS}]$

Table 5.2.1. Linear fit slopes of features as a function of percent evaporator air flow rate fault at an indoor dry-bulb temperature of 21.1 °C and outdoor conditions of -8.3 °C/Dry

Feature Name	Feature Symbol	Feature's slope	Feature's % Change @ Max Fault Level
total air side capacity (kW %$^{-1}$)	Q_{CA}	-4.10E-02	-23.7
compressor power (kW %$^{-1}$)	W_{comp}	-2.96E-03	-5.7
refrigerant mass flow rate (kg min^{-1} %$^{-1}$)	m_R	-1.29E-02	-27.6
coefficient of performance(%-1)	COP	-1.80E-02	-23.5
indoor unit refrigerant side capacity (kW %$^{-1}$)	Q_{CR}	-4.38E-02	-26.4

Outdoor coil face area blocked to produce higher air pressure drop across the coil. Fault determined by % coil face area blocked equal to 10, 20, and 30 percent.	Max Fault Level: 30.0 %

Listed in Descending Order of Largest ABS(Δ°C (%Fault)$^{-1}$)

		Residual's slope as a function of % fault level	
		Δ°C (%Fault)$^{-1}$	Feature % Change @ Max Fault Level
evaporator exit saturation temperature (°C)	T_{ER}	-0.261	-339.2
evaporator exit superheat (°C)	ΔT_{shE}	0.250	128.2
reversing valve temperature change, suction side (°C)	ΔT_{RVS}	-0.248	-351.0
evaporator bend thermocouple, TC#103 (°C)	T_{E103}	-0.188	-122.4
outdoor temperature minus TC#103 (°C)	ΔT_{103}	0.187	116.0
reversing valve temperature change, discharge side (°C)	ΔT_{RVD}	-0.180	-284.6
compressor discharge wall temperature (°C)	T_D	0.138	4.8
condenser bend thermocouple, TC#15 (°C)	T_{C15}	-0.104	-6.4
condenser inlet saturation temperature (°C)	T_{CR}	-0.097	-6.0
ID fan motor case temperature (°C)	T_{IDF}	-0.085	-3.3
condenser air temperature rise (°C)	ΔT_{CA}	-0.073	-23.1
vapor superheat at outdoor service valve (°C)	ΔT_{shV}	0.058	5.2
evaporator air temperature drop (°C)	ΔT_{EA}	-0.042	-45.5
liquid line temperature drop (°C)	ΔT_{LL}	0.027	19.5
liquid line subcooling at outdoor service valve (°C)	ΔT_{scV}	0.023	10.5
OD fan motor case temperature (°C)	T_{ODF}	-0.010	-0.8
condenser inlet superheat (°C)	ΔT_{shC}	0.000	-0.7

5.2.2   Indoor dry-bulb temperature of 21.1 °C at outdoor conditions of 8.3 °C/72.5 % RH

Figure 5.2.6 shows the change in air-side heating capacity, refrigerant-side heating capacity, compressor power, COP, and refrigerant mass flow rate as a function of the percent area blockage of the outdoor coil face. COP was calculated based upon refrigerant-side heating capacity. Heating capacity and COP dropped by 18 % and 14 %, respectively, at the maximum fault level of 30.0 % outdoor coil area blockage.

Figure 5.2.7 shows the residuals of T_{ER}, T_{E103}, and ΔT_{shE} as a function of indoor coil air flow rate percent reduction. Evaporator exit superheat residual and evaporator exit saturation temperature residual show comparable changes with outdoor coil area blockage.

Figure 5.2.8 shows residuals for T_D, T_{CR}, T_{C15}, ΔT_{shC}, ΔT_{shV}, and ΔT_{scV}. The residuals of condenser inlet refrigerant saturation temperature, T_{CR}, its twin, T_{C15}, and compressor discharge line wall temperature, T_D, showed substantial negative slopes.

Figure 5.2.9 shows residuals for ΔT_{CA}, ΔT_{EA}, ΔT_{LL}, and ΔT_{103}. The residual of the temperature difference feature (outdoor air temperature minus evaporator coil return bend temperature, TC#103, showed a large positive slope for this fault. The residual of the temperature rise across the indoor coil, ΔT_{CA}, showed the second largest slope among these features.

Figure 5.2.10 shows residuals for T_{IDF}, T_{ODF}, ΔT_{RVD}, and ΔT_{RVS}. The indoor fan motor case temperature, T_{IDF}, and the reversing valve refrigerant temperature change on the suction side, ΔT_{RVS}, showed big residual slopes that could be useful for FDD.

Table 5.2.2 shows the linear slopes and absolute value of the percent changes in the system characteristics and residual temperatures for the outdoor coil area blockage fault. Four features had residual slopes greater than 0.2 C %$^{-1}$; ΔT_{RVS}, T_{ER}, ΔT_{shE}, and T_{IDF}. Of these four features, ΔT_{RVS} decreased by 642 % from its NFSS value at the maximum fault level of 30.0 %. Evaporator exit superheat increased by 97 % from 5.5 C to 9.6 C.

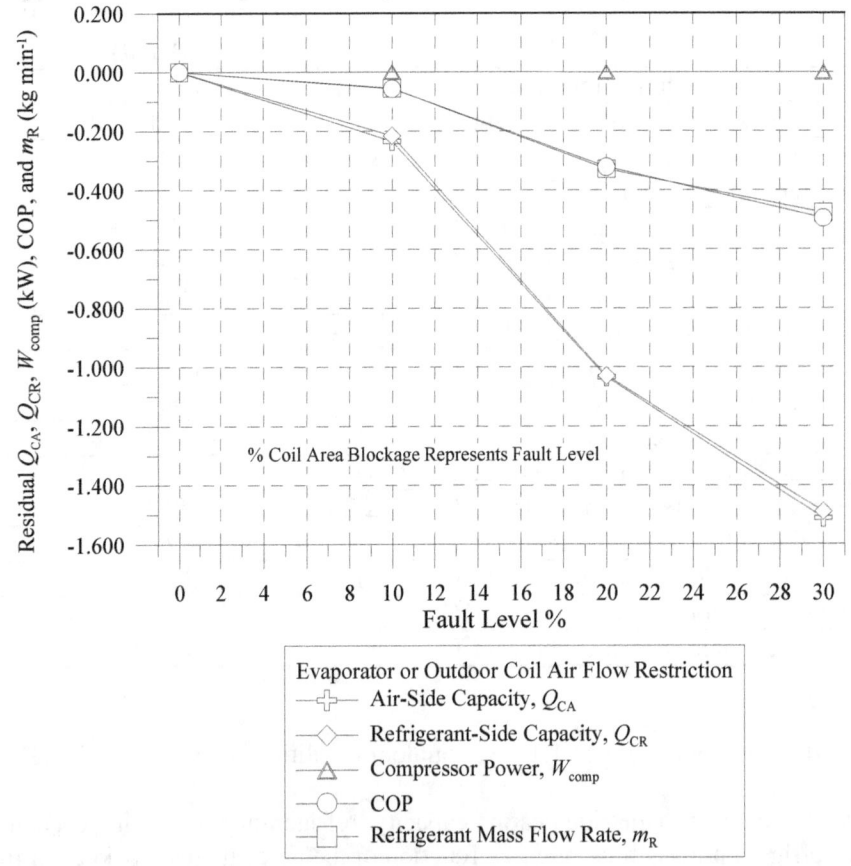

Figure 5.2.6. Residual of selected features with evaporator air flow faults imposed at an indoor dry-bulb temperature of 21.1 °C and outdoor conditions of 8.3 °C/72.5 % RH: R[Q_{CA}], R[Q_{CR}], R[W_{comp}], R[COP], and R[m_R]

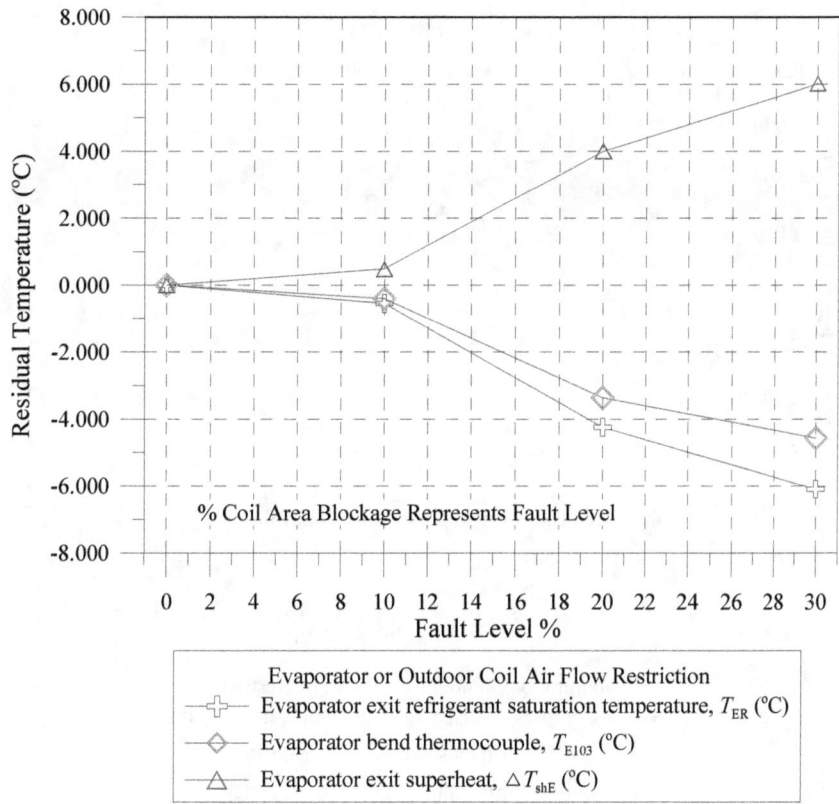

Figure 5.2.7. Residual of selected features with an evaporator fouling fault at an indoor dry-bulb temperature of 21.1 °C and outdoor conditions of 8.3 °C/72.5 % RH: R[T_{ER}], R[T_{E103}], and R[ΔT_{shE}]

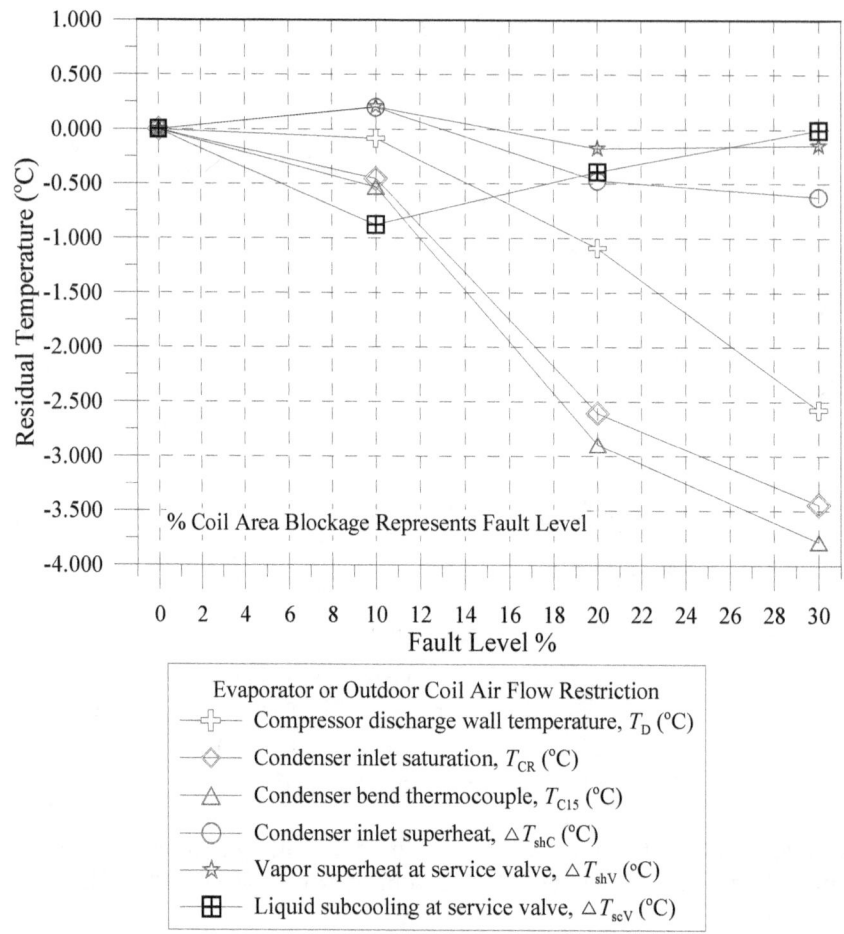

Figure 5.2.8. Residual of selected features with an evaporator fouling fault at an indoor dry-bulb temperature of 21.1 °C and outdoor conditions of 8.3 °C/72.5 % RH: $R[T_D]$, $R[T_{CR}]$, $R[T_{C15}]$, $R[\Delta T_{shC}]$, $R[\Delta T_{shV}]$, and $R[\Delta T_{scV}]$

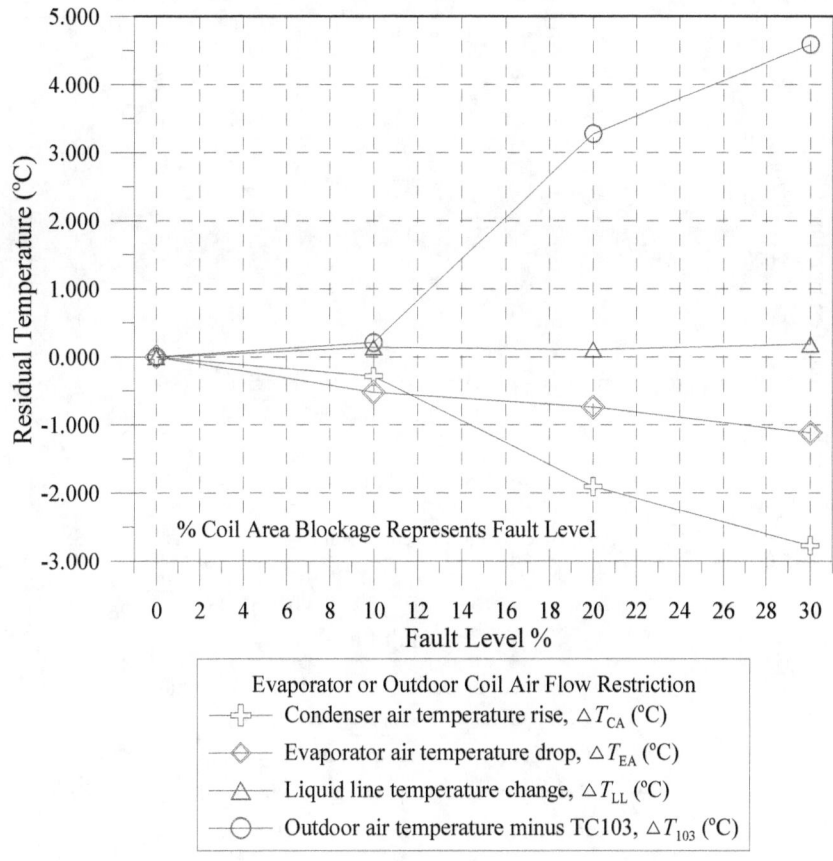

Figure 5.2.9. Residual of selected features with an evaporator fouling fault at an indoor dry-bulb
temperature of 21.1 °C and outdoor conditions of 8.3 °C/72.5 % RH: R[ΔT_{CA}], R[ΔT_{EA}],
R[ΔT_{LL}], and R[ΔT_{103}]

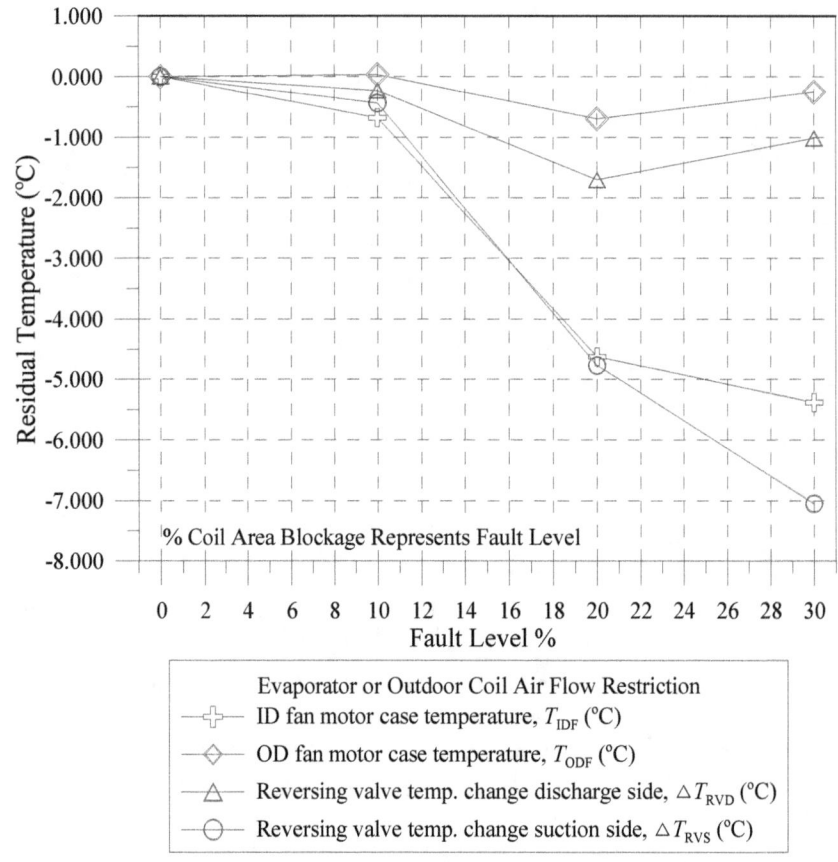

Figure 5.2.10. Residual of selected features with an evaporator fouling fault at an indoor dry-bulb temperature of 21.1 °C and outdoor conditions of 8.3 °C/72.5 % RH: R[T_{IDF}], R[T_{ODF}], R[ΔT_{RVD}], and R[ΔT_{RVS}]

Table 5.2.2. Linear fit slopes of features as a function of percent evaporator air flow rate fault for a given nominal test conditions of 21.1 °C indoor and 8.3 °C/72.5 % RH outdoor

Feature Name	Feature Symbol	Feature's slope	Feature % Change @ Max Fault Level
total air side capacity (kW %$^{-1}$)	Q_{CA}	-5.34E-02	-18.1
compressor power (kW %$^{-1}$)	W_{comp}	-4.02E-03	-6.1
refrigerant mass flow rate (kg min^{-1} %$^{-1}$)	m_R	-1.70E-02	-19.4
coefficient of performance(%-1)	COP	-1.75E-02	-14.2
indoor unit refrigerant side capacity (kW %$^{-1}$)	Q_{CR}	-5.28E-02	-18.0

Outdoor coil face area blocked to produce lower air flow rate and higher air pressure drop across the coil. Fault determined by % coil face area blocked equal to 10, 20, and 30 percent.

Max Fault Level: 30.0 %

Listed in Descending Order of Largest ABS(ΔT°C (%Fault)$^{-1}$)

		Residual's slope as a function of % fault level	
		Δ°C (% Fault)$^{-1}$	Feature % Change @ Max Fault Level
reversing valve temperature change, suction side (°C)	ΔT_{RVS}	-0.255	-642.9
evaporator exit saturation temperature (°C)	T_{ER}	-0.220	-33.6
evaporator exit superheat (°C)	ΔT_{shE}	0.216	97.9
ID fan motor case temperature (°C)	T_{IDF}	-0.201	-6.3
outdoor temperature minus TC#103 (°C)	ΔT_{103}	0.168	72.7
evaporator bend thermocouple, TC#103 (°C)	T_{E103}	-0.167	-23.0
condenser bend thermocouple, TC#15 (°C)	T_{C15}	-0.137	-6.6
condenser inlet saturation temperature (°C)	T_{CR}	-0.125	-6.1
condenser air temperature rise (°C)	ΔT_{CA}	-0.099	-18.5
compressor discharge wall temperature (°C)	T_D	-0.087	-3.3
reversing valve temperature change, discharge side (°C)	ΔT_{RVD}	-0.045	-31.8
evaporator air temperature drop (°C)	ΔT_{EA}	-0.036	-27.8
condenser inlet superheat (°C)	ΔT_{shC}	-0.025	-2.7
OD fan motor case temperature (°C)	T_{ODF}	-0.015	-0.5
vapor superheat at outdoor service valve (°C)	ΔT_{shV}	-0.008	-0.6
liquid line temperature drop (°C)	ΔT_{LL}	0.005	8.4
liquid line subcooling at outdoor service valve (°C)	ΔT_{scV}	0.005	-0.2

5.3 Compressor or Four-Way Valve Leakage Fault (CMF fault)

5.3.1   Indoor dry-bulb temperature of 21.1 °C at outdoor conditions of -8.3 °C/Dry

The compressor or four-way valve refrigerant leakage fault was simulated by bypassing hot discharge gas to the compressor suction, thus decreasing refrigerant mass flow rate. Figure 5.3.1 shows the change in heating capacity, refrigerant-side heating capacity, compressor power, COP, and refrigerant mass flow rate as a function of the percent reduction in refrigerant mass flow rate. Refrigerant-side heating capacity and COP decreased by 16.6 % and 13.8 %, respectively, with a 17.5 % reduction in refrigerant mass flow rate.

Figure 5.3.2 shows the residuals of T_{ER}, T_{E103}, and ΔT_{shE} as a function of percent decrease in refrigerant mass flow rate due to the hot gas bypass. At the maximum fault level, the evaporator exit superheat, ΔT_{shE}, and evaporator exit refrigerant saturation temperature, T_{ER}, changed by more than 5 % from their NFSS values.

Figure 5.3.3 shows residuals for T_D, T_{CR}, T_{C15}, ΔT_{shC}, ΔT_{shV}, and ΔT_{scV}. The key feature to observe in this plot is the negative slope of the condenser inlet saturation temperature as shown by the two features T_{CR} and T_{C15}. The residuals of these two features had the greatest change of all residuals and are thus the strongest indicators of a compressor or four-way valve leakage fault. Both features had a greater than 0.1 °C change with every percent change in refrigerant hot gas bypass mass flow rate.

Figure 5.3.4 shows residuals for ΔT_{CA}, ΔT_{EA}, ΔT_{LL}, and ΔT_{103}. Of these features, the evaporator air temperature change had the largest percent change at the highest fault level. The residuals of all of these features showed changes as fault level increased; the maximum change was seen for the condenser air temperature, ΔT_{CA}.

Figure 5.3.5 shows residuals for T_{IDF}, T_{ODF}, ΔT_{RVD}, and ΔT_{RVS}. Compared to the NFSS value, the temperature change across the discharge side of the four-way valve showed an immediate increase with the addition of hot gas bypass, then decreased with increasing fault level, and returned to normal levels at a fault level near 13 % with a further decrease at the maximum fault level. The indoor fan motor case temperature also showed a decreasing residual with increasing fault level.

Table 5.3.1 shows the residual's linear slopes and the percent changes in the system characteristics and temperatures for the compressor or four-way valve leakage fault. This fault did not produce large changes in residuals as fault level increased. The greatest change in residual occurred for the condenser inlet refrigerant saturation temperature, as indicated by T_{CR} and T_{C15}. Several features had large changes in value; T_{ER}, ΔT_{EA}, ΔT_{RVD}, T_{E103}, ΔT_{LL}, and ΔT_{103} changed by more than 10 % from their NFSS values, but the change in their residual values with respect to increasing fault level was small.

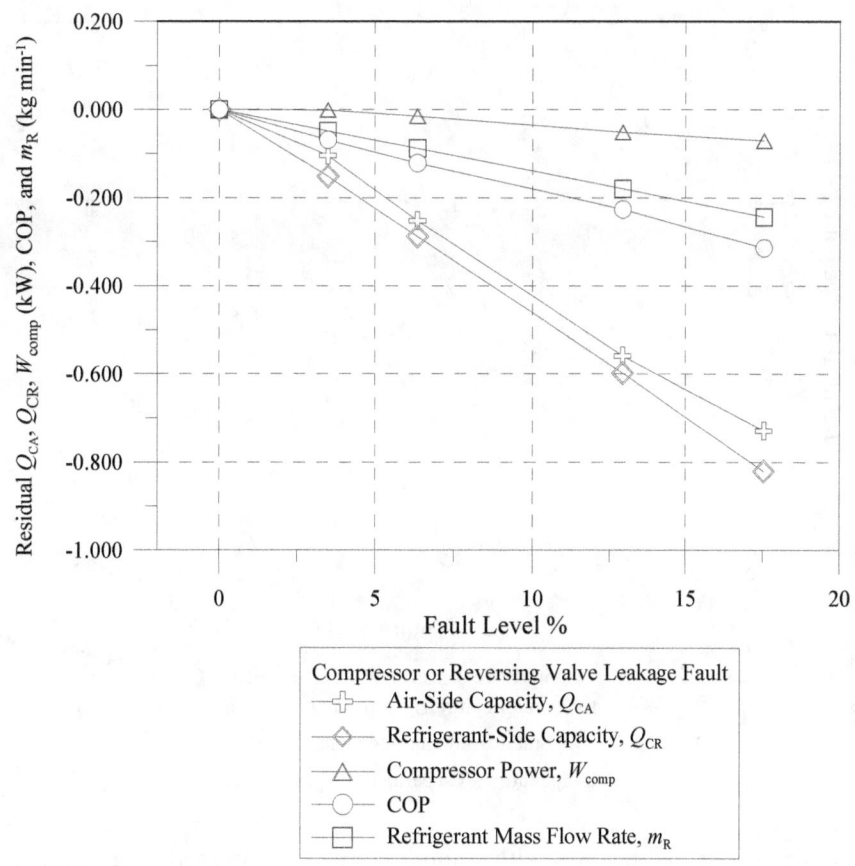

Figure 5.3.1. Residual of selected features with compressor or four-way valve refrigerant leakage faults imposed at an indoor dry-bulb temperature of 21.1 °C and outdoor conditions of -8.3 °C/Dry: R[Q_{CA}], R[Q_{CR}], R[W_{comp}], R[COP], and R[m_R]

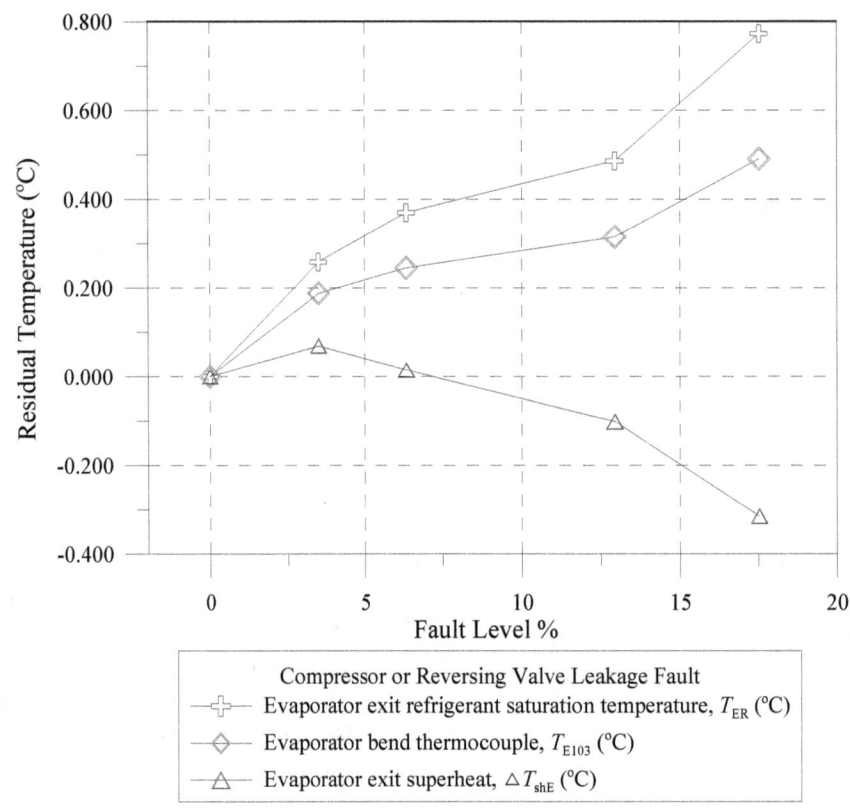

Figure 5.3.2. Residual of selected features with compressor or four-way valve refrigerant leakage fault at an indoor dry-bulb temperature of 21.1 °C and outdoor conditions of -8.3 °C/Dry: R[T_{ER}], R[T_{E103}], and R[ΔT_{shE}]

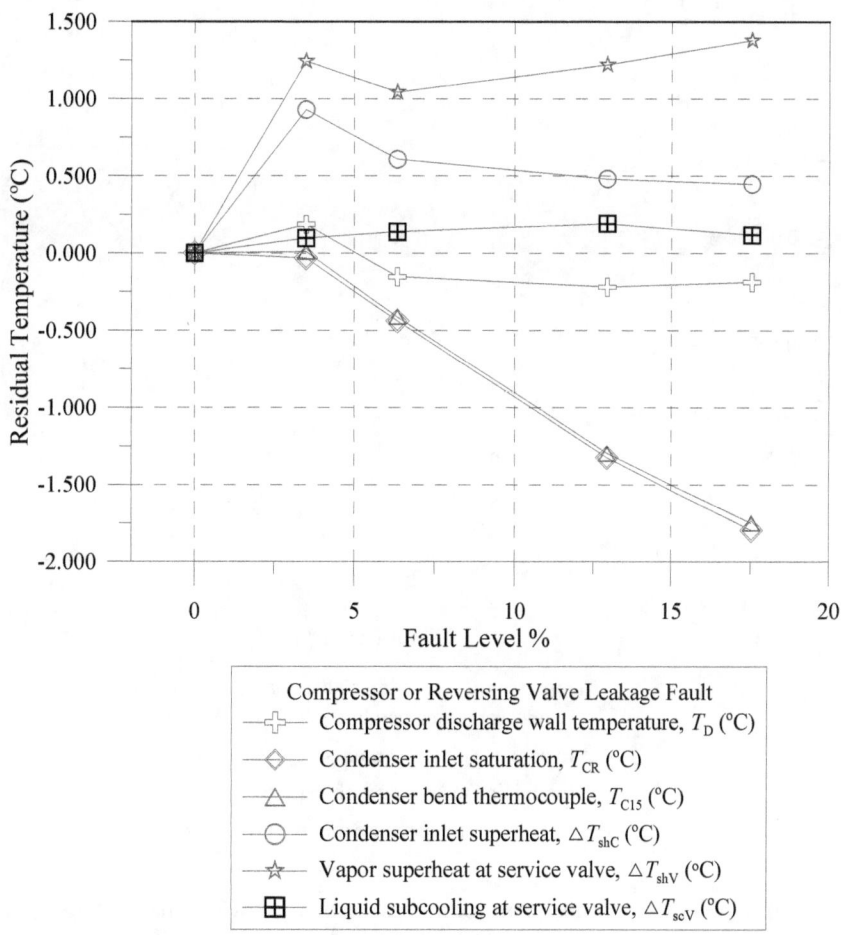

Figure 5.3.3. Residual of selected features with compressor or four-way valve refrigerant leakage fault at an indoor dry-bulb temperature of 21.1 °C and outdoor conditions of -8.3 °C/Dry: R[T_D], R[T_{CR}], R[T_{C15}], R[ΔT_{shC}], R[ΔT_{shV}], and R[ΔT_{scV}]

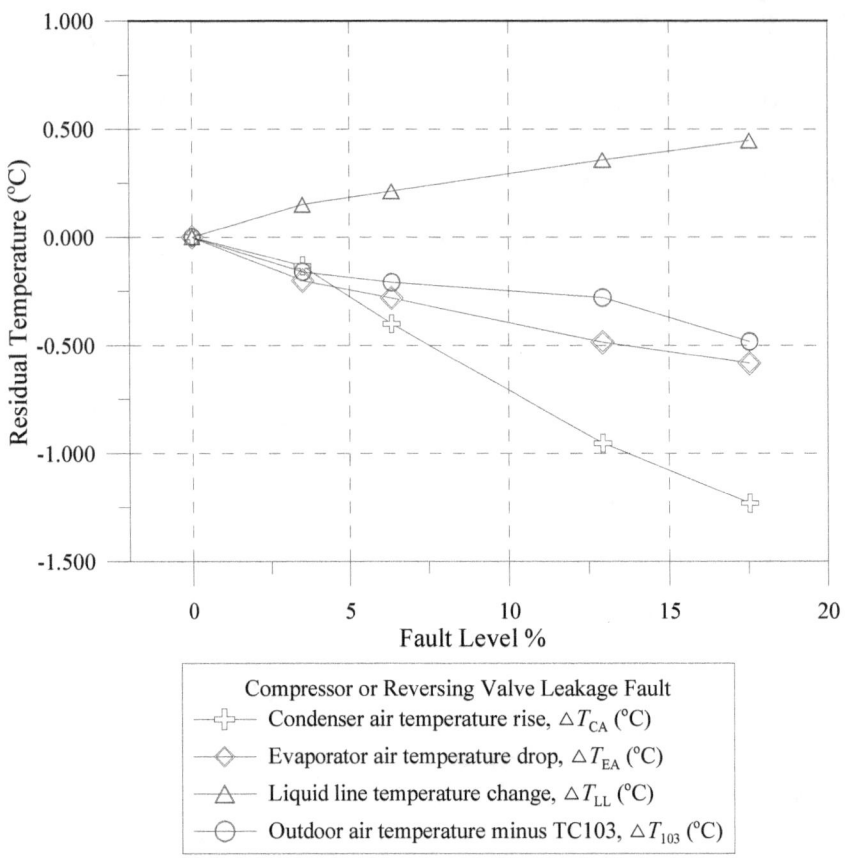

Figure 5.3.4. Residual of selected features with compressor or four-way valve refrigerant leakage fault at an indoor dry-bulb temperature of 21.1 °C and outdoor conditions of -8.3 °C/Dry: R[ΔT_{CA}], R[ΔT_{EA}], R[ΔT_{LL}], and R[ΔT_{103}]

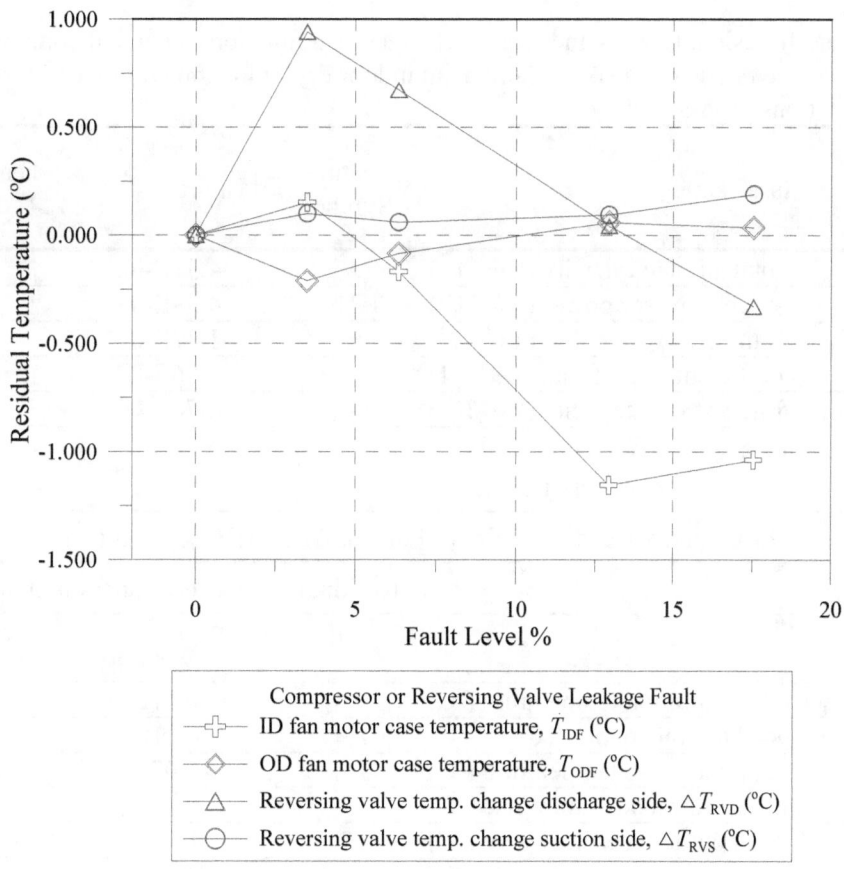

Figure 5.3.5. Residual of selected features with compressor or four-way valve refrigerant leakage fault at an indoor dry-bulb temperature of 21.1 °C and outdoor conditions of -8.3 °C/Dry: $R[T_{IDF}]$, $R[T_{ODF}]$, $R[\Delta T_{RVD}]$, and $R[\Delta T_{RVS}]$

Table 5.3.1. Linear fit residual slopes and feature changes as a function of percent compressor or four-way valve refrigerant leakage fault at an indoor dry-bulb temperature of 21.1 °C and outdoor conditions of -8.3 °C/Dry

Feature Name	Feature Symbol	Feature's slope	Feature's % Change @ Max Fault Level
total air side capacity (kW %$^{-1}$)	Q_{CA}	-4.31E-02	-14.2
compressor power (kW %$^{-1}$)	W_{comp}	-4.39E-03	-4.4
refrigerant mass flow rate (kg min^{-1} %$^{-1}$)	m_R	-1.39E-02	-17.5
coefficient of performance(%-1)	COP	-1.76E-02	-13.8
indoor unit refrigerant side capacity (kW %$^{-1}$)	Q_{CR}	-4.70E-02	-16.6

Compressor hot gas bypassed to suction. Fault determined by % decrease in refrigerant mass flow rate.

Max Fault Level: 17.5 %

Listed in Descending Order of Largest ABS(ΔT°C (% Fault)$^{-1}$)

		Residual's slope as a function of % fault level	
		Δ°C (% Fault)$^{-1}$	Feature % Change @ Max Fault Level
condenser inlet saturation temperature (°C)	T_{CR}	-0.112	-3.6
condenser bend thermocouple, TC#15 (°C)	T_{C15}	-0.110	-3.5
ID fan motor case temperature (°C)	T_{IDF}	-0.078	-1.3
condenser air temperature rise (°C)	ΔT_{CA}	-0.074	-13.2
vapor superheat at outdoor service valve (°C)	ΔT_{shV}	0.057	4.6
reversing valve temperature change, discharge side (°C)	ΔT_{RVD}	-0.041	-18.1
evaporator exit saturation temperature (°C)	T_{ER}	0.039	33.5
evaporator air temperature drop (°C)	ΔT_{EA}	-0.032	-21.2
liquid line temperature drop (°C)	ΔT_{LL}	0.024	10.6
evaporator bend thermocouple, TC#103 (°C)	T_{E103}	0.024	10.7
outdoor temperature minus TC#103 (°C)	ΔT_{103}	-0.024	-10.0
evaporator exit superheat (°C)	ΔT_{shE}	-0.019	-5.3
compressor discharge wall temperature (°C)	T_D	-0.017	-0.2
OD fan motor case temperature (°C)	T_{ODF}	0.009	0.1
reversing valve temperature change, suction side (°C)	ΔT_{RVS}	0.008	9.2
liquid line subcooling at outdoor service valve (°C)	ΔT_{scV}	0.007	1.8
condenser inlet superheat (°C)	ΔT_{shC}	0.006	1.8

5.3.2   Indoor dry-bulb temperature of 21.1 °C at outdoor conditions of 8.3 °C/72.5 % RH

Figure 5.3.6 shows the change in heating capacity, refrigerant-side heating capacity, compressor power, COP, and refrigerant mass flow rate as a function of the percent reduction refrigerant mass flow rate due to a compressor or four-way valve refrigerant leakage fault. Heating capacity and COP dropped by 11.6 % and 7.7 %, respectively, at the maximum fault level of 12.4 %.

Figure 5.3.7 shows the residuals of T_{ER}, T_{E103}, and ΔT_{shE} as a function of percent decrease in refrigerant mass flow rate due to the hot gas bypass. These features showed minimal changes with this fault.

Figure 5.3.8 shows residuals for T_D, T_{CR}, T_{C15}, T_{shC}, T_{shV}, and T_{scV}. The residuals of the vapor line superheat and condenser inlet superheat showed almost no change until fault levels of 5 % and higher. Liquid line subcooling oscillated around its NFSS value as fault level increased. The residuals of compressor discharge line wall temperature, T_D, and condenser inlet saturation temperature, T_{CR}, showed distinct negative slopes with increasing fault level.

Figure 5.3.9 shows residuals for ΔT_{CA}, ΔT_{EA}, ΔT_{LL}, and ΔT_{103}. The residual of the condenser air temperature rise had the greatest slope of these feature residuals with a value of -0.12 C %$^{-1}$. The other three residuals vary by less than 0.5 °C from their NFSS values at the maximum fault level.

Figure 5.3.10 shows residuals for T_{IDF}, T_{ODF}, ΔT_{RVD}, and ΔT_{RVS}. The indoor fan motor case temperature showed the largest change of these features with a residual slope of -0.13 C %$^{-1}$. The remaining residuals varied by less than 0.5 °C from their NFSS values.

Table 5.3.2 shows the linear slopes and absolute value of the percent changes in the system characteristics and residual temperatures for the compressor or four-way valve leakage fault. Four features had residual slopes greater than 0.12 C %$^{-1}$; T_{CR}, T_{C15}, T_{IDF}, and ΔT_{CA}. Of these four features, ΔT_{CA} decreased by 11 % from its NFSS value at the maximum fault level of 12.4 %.

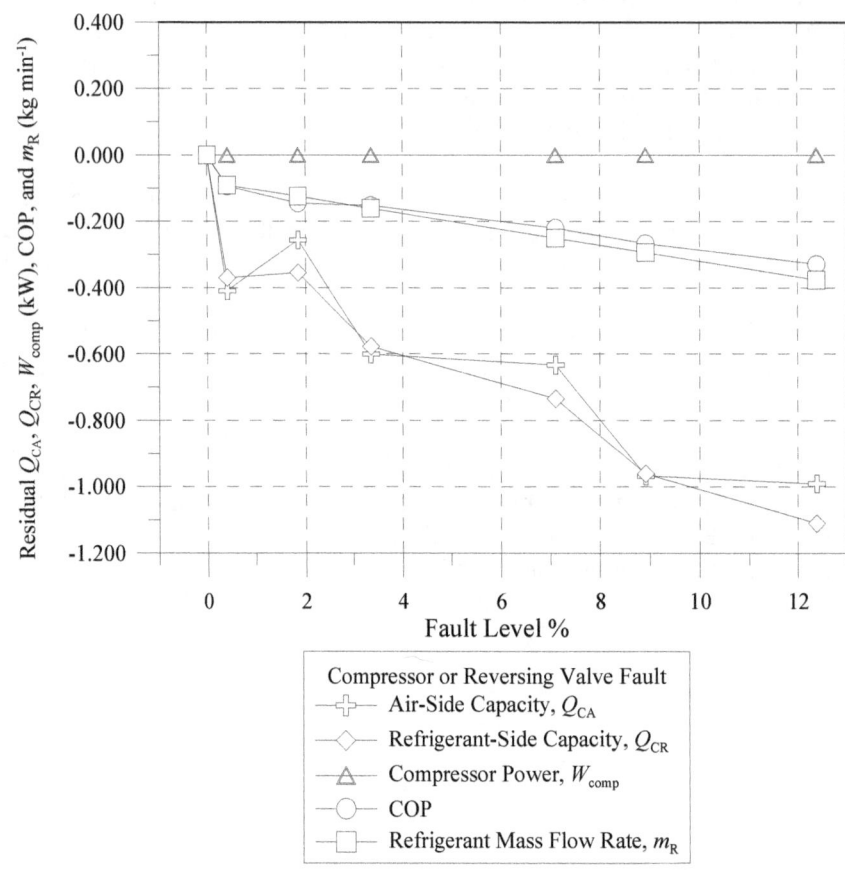

Figure 5.3.6. Residual of selected features with compressor or four-way valve refrigerant leakage faults imposed at an indoor dry-bulb temperature of 21.1 °C and outdoor conditions of 8.3 °C/72.5 % RH: R[Q_{CA}], R[Q_{CR}], R[W_{comp}], R[COP], and R[m_R]

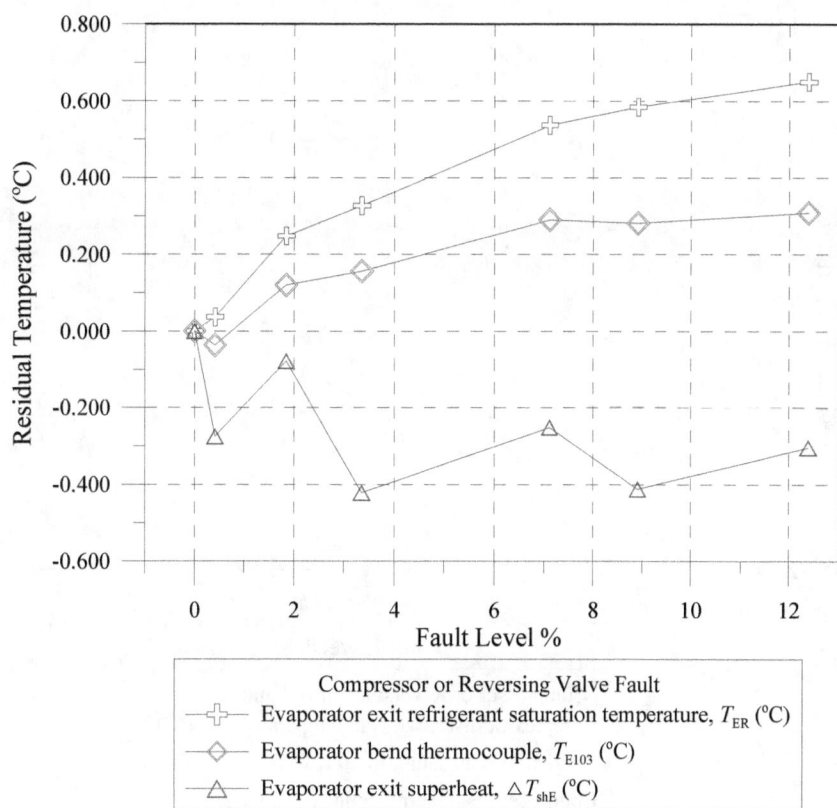

Figure 5.3.7. Residual of selected features with compressor or four-way valve refrigerant leakage fault at an indoor dry-bulb temperature of 21.1 °C and outdoor conditions of 8.3 °C/72.5 % RH: $R[T_{ER}]$, $R[T_{E103}]$, and $R[\Delta T_{shE}]$

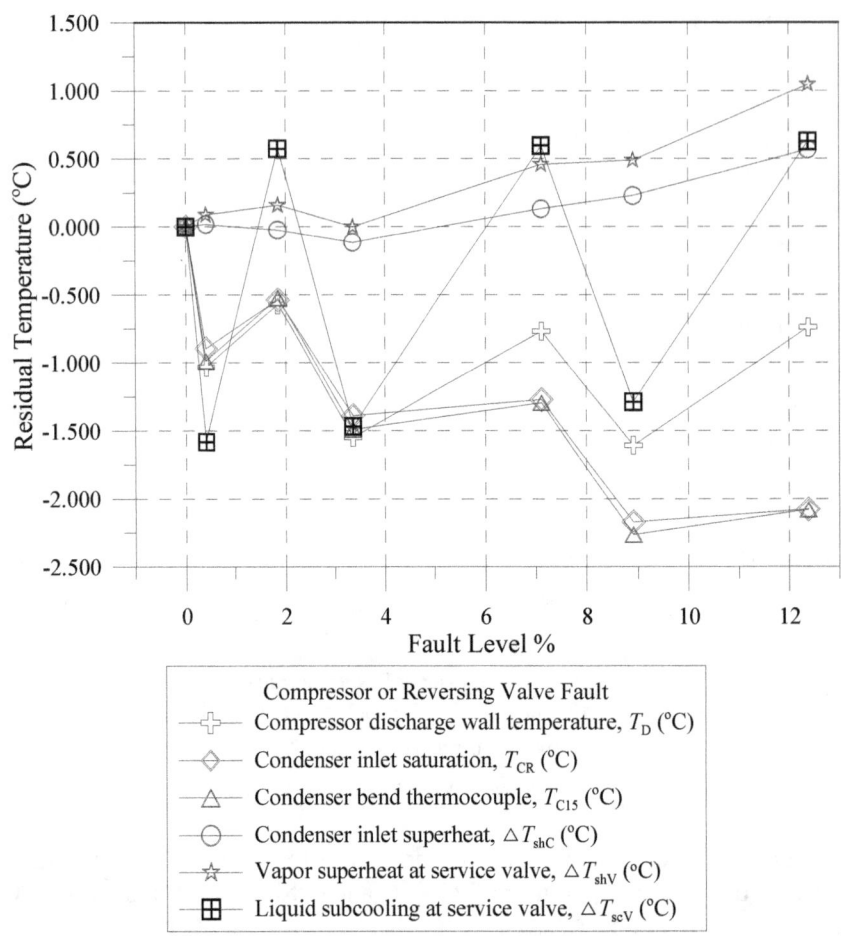

Figure 5.3.8. Residual of selected features with compressor or four-way valve refrigerant leakage fault: R[T_D], R[T_{CR}], R[T_{C15}], R[ΔT_{shC}], R[ΔT_{shV}], and R[ΔT_{scV}] at an indoor dry-bulb temperature of 21.1 °C and outdoor conditions of 8.3 °C/72.5 % RH

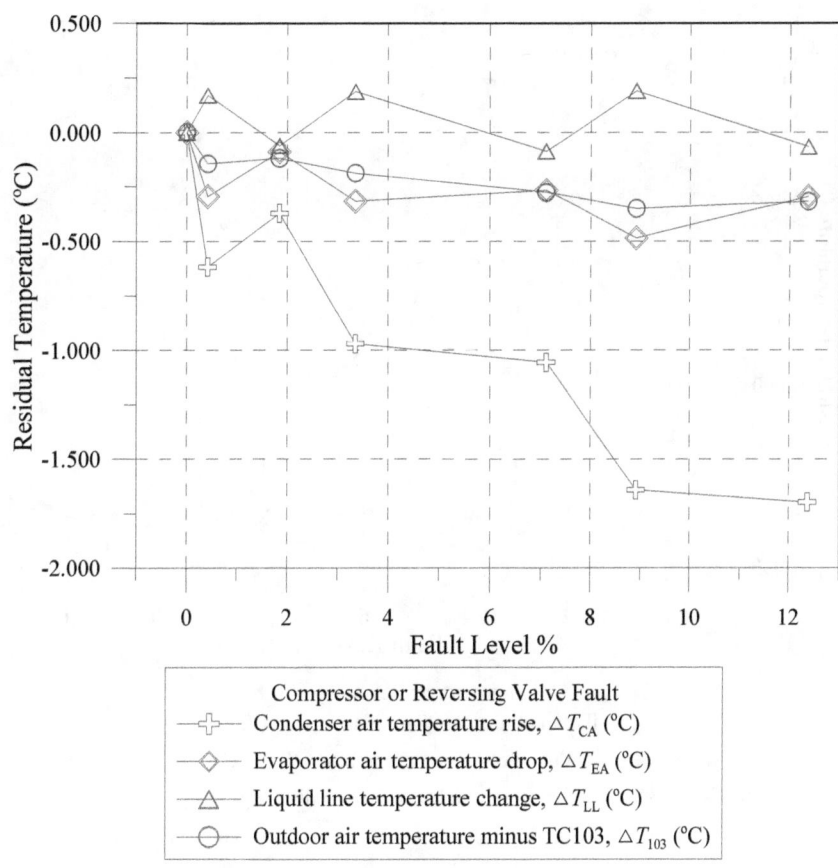

Figure 5.3.9. Residual of selected features with compressor or four-way valve refrigerant leakage fault at an indoor dry-bulb temperature of 21.1 °C and outdoor conditions of 8.3 °C/72.5 % RH: R[ΔT_{CA}], R[ΔT_{EA}], R[ΔT_{LL}], and R[ΔT_{103}]

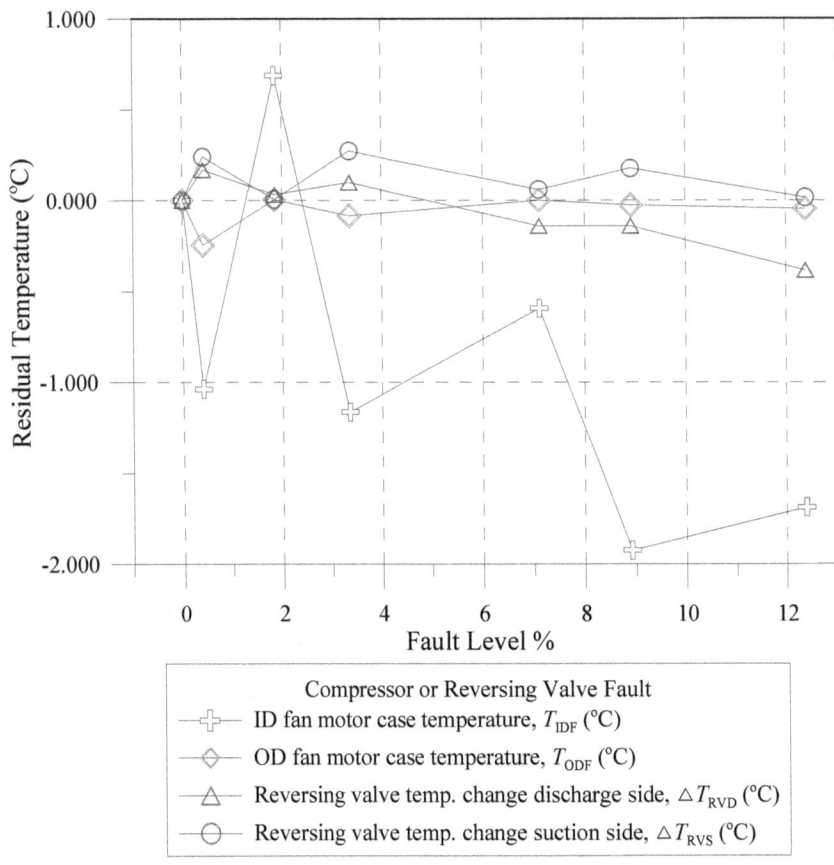

Figure 5.3.10. Residual of selected features with compressor or four-way valve refrigerant leakage fault at an indoor dry-bulb temperature of 21.1 °C and outdoor conditions of 8.3 °C/72.5 % RH: R[T_{IDF}], R[T_{ODF}], R[ΔT_{RVD}], and R[ΔT_{RVS}]

Table 5.3.2. Linear fit slopes of features as a function of percent compressor or four-way valve leakage fault at nominal test conditions of 21.1 °C indoor and 8.3 °C/72.5 % RH outdoor

Feature Name	Feature Symbol	Feature's slope	Feature % Change @ Max Fault Level
total air side capacity (kW %$^{-1}$)	Q_{CA}	-7.01E-02	-11.6
compressor power (kW %$^{-1}$)	W_{comp}	-7.10E-03	-5.5
refrigerant mass flow rate (kg min^{-1} %$^{-1}$)	m_R	-2.68E-02	-11.9
coefficient of performance(%$^{-1}$)	COP	-2.23E-02	-7.7
indoor unit refrigerant side capacity (kW %$^{-1}$)	Q_{CR}	-7.77E-02	-11.6

Compressor hot gas bypassed to suction. Fault determined by % decrease in refrigerant mass flow rate.	Max Fault Level: 12.4 %

Listed in Descending Order of Largest ABS(ΔT°C (% Fault)$^{-1}$)

		Residual's slope as a function of % fault level	
		Δ°C (% Fault)$^{-1}$	Feature % Change @ Max Fault Level
condenser inlet saturation temperature (°C)	T_{CR}	-0.148	-3.9
condenser bend thermocouple, TC#15 (°C)	T_{C15}	-0.147	-4.0
ID fan motor case temperature (°C)	T_{IDF}	-0.132	-2.3
condenser air temperature rise (°C)	ΔT_{CA}	-0.124	-10.9
vapor superheat at outdoor service valve (°C)	ΔT_{shV}	0.075	1.9
liquid line subcooling at outdoor service valve (°C)	ΔT_{scV}	0.063	-31.8
evaporator exit saturation temperature (°C)	T_{ER}	0.053	3.2
condenser inlet superheat (°C)	ΔT_{shC}	0.043	1.0
compressor discharge wall temperature (°C)	T_D	-0.039	-2.1
reversing valve temperature change, discharge side (°C)	ΔT_{RVD}	-0.036	-4.4
evaporator bend thermocouple, TC#103 (°C)	T_{E103}	0.028	1.4
outdoor temperature minus TC#103 (°C)	ΔT_{103}	-0.024	-5.6
evaporator air temperature drop (°C)	ΔT_{EA}	-0.021	-12.1
evaporator exit superheat (°C)	ΔT_{shE}	-0.018	-6.7
liquid line temperature drop (°C)	ΔT_{LL}	-0.006	8.5
OD fan motor case temperature (°C)	T_{ODF}	0.006	-0.1
reversing valve temperature change, suction side (°C)	ΔT_{RVS}	-0.004	16.1

5.4 Liquid Line Restriction Fault (LL fault)

5.4.1   Indoor dry-bulb temperature of 21.1 °C at outdoor conditions of -8.3 °C/Dry

The refrigerant liquid line restriction fault was simulated by partially closing a pair of parallel valves near the middle of the liquid line. Figure 5.4.1 shows the change in air-side heating capacity, refrigerant-side heating capacity, compressor power, COP, and refrigerant mass flow rate as a function of the percent increase in refrigerant pressure drop. Refrigerant-side heating capacity and COP changed by only 0.03 % and 0.12 %, respectively, with a 45.4 % increase in refrigerant pressure drop. NFSS pressure drop was 33 kPa and increased to 48 kPa at the highest fault level; compressor discharge pressure increased less than 15 kPa. Suction pressure decreased by less than 10 kPa at the highest fault level.

Figure 5.4.2 shows the residuals of T_{ER}, T_{E103}, and ΔT_{shE} as a function of liquid line refrigerant pressure drop. The residual of the evaporator exit superheat, ΔT_{shE}, oscillated due to corrective actions by the thermostatic expansion valve. At the maximum fault level ΔT_{shE} was within 0.1 °C of its NFSS value. The pressure calculated value of the evaporator exit saturation temperature, T_{ER}, and the thermocouple measured value, T_{E103}, decreased with increasing fault level, but the magnitude of the residual value slopes was less than 0.01 °C %$^{-1}$.

Figure 5.4.3 shows residuals for T_D, T_{CR}, T_{C15}, ΔT_{shC}, ΔT_{shV}, and ΔT_{scV}. The results of the thermostatic expansion valve's corrective actions are seen within this plot. There is a linear change in superheat values and compressor discharge wall temperatures, and then a correction by the TXV occurred at a fault level somewhere between 20 % and 35 %. As the fault level increased above 35 %, the beginning of a new linear trend may be occurring.

Figure 5.4.4 shows residuals for ΔT_{CA}, ΔT_{EA}, ΔT_{LL}, and ΔT_{103}. Of these features, the liquid line temperature change, ΔT_{LL}, had the largest percent change at the highest fault level, a decrease of 8.03 %. The residuals of all of these features showed changes as fault level increased; the maximum was seen for the condenser air temperature change with a residual slope of -0.074 °C %$^{-1}$. The rapid drop of liquid line temperature change residual at the 8 % fault level did not continue as fault level increased. This suggests that an offset from the NFSS value was already present, or there was a very non-linear effect occurring with a small increase in liquid line pressure drop.

Figure 5.4.5 shows residuals for T_{IDF}, T_{ODF}, ΔT_{RVD}, and ΔT_{RVS}. Compared to the NFSS value, the temperature change across the cold side of the four-way valve showed an immediate increase with increased liquid line pressure drop, and then decreased with increasing fault level. The indoor fan motor case temperature also showed an increasing residual with increasing fault level, while the outdoor fan motor case temperature was decreasing; they appear to be almost mirror images about the zero residual x-axis.

Table 5.4.1 shows the residual's linear slopes and the percent changes in the system characteristics and temperatures for the liquid line pressure drop fault. This fault did not produce large changes in residuals as fault level increased. The greatest change in residual (greatest slope) occurred for the refrigerant subcooling at the service valve; its value was -0.014 °C %$^{-1}$. Several features had large changes in value at the 45 % maximum fault level; T_{ER}, ΔT_{RVS}, and ΔT_{scV} changed by more than 10 % from their NFSS values, but the change in their residual values with respect to increasing fault level was small (less than 0.02 °C %$^{-1}$). All of the figures show that the liquid line refrigerant flow restriction fault has a minimal effect on the system performance at these temperature conditions due to the corrective action of the TXV, and this fault is not evident from changes in the features listed in Table 5.4.1. The high fault level is not indicative of the small absolute change in liquid line pressure drop; low mass flow rates a low ambient conditions produce low pressure drop in the liquid line and thus even a small absolute change in pressure drop will produce a large percentage change.

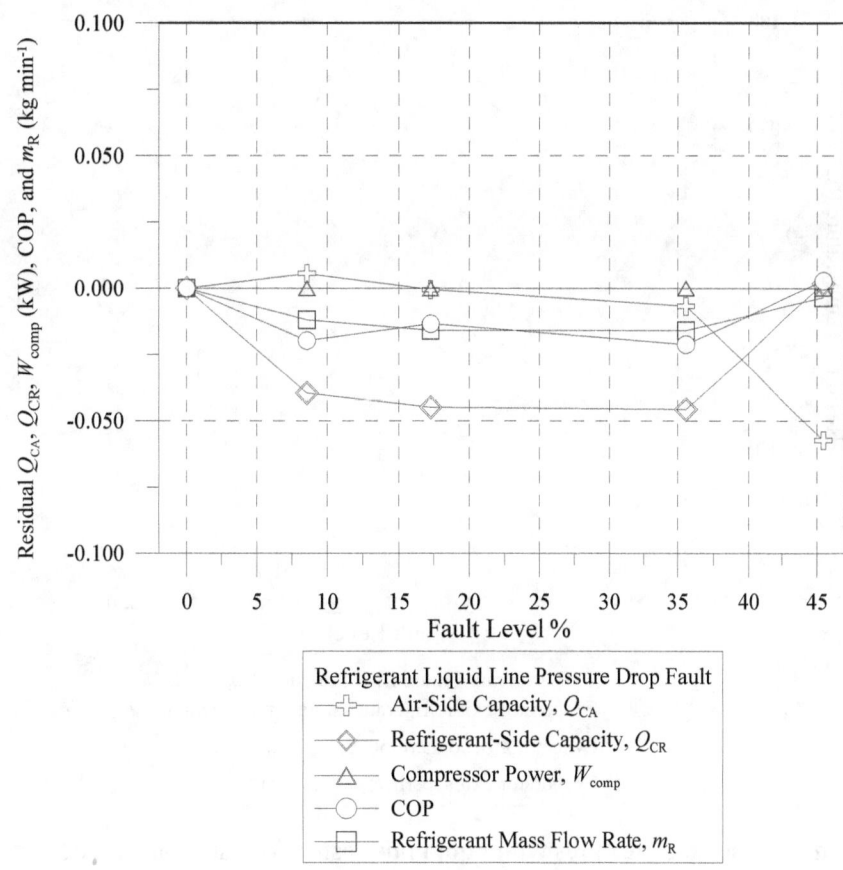

Figure 5.4.1. Residual air-side capacity, refrigerant-side capacity, compressor power, COP, and refrigerant mass flowrate with liquid line restriction faults imposed at an indoor dry-bulb temperature of 21.1 °C and outdoor conditions of -8.3 °C/Dry

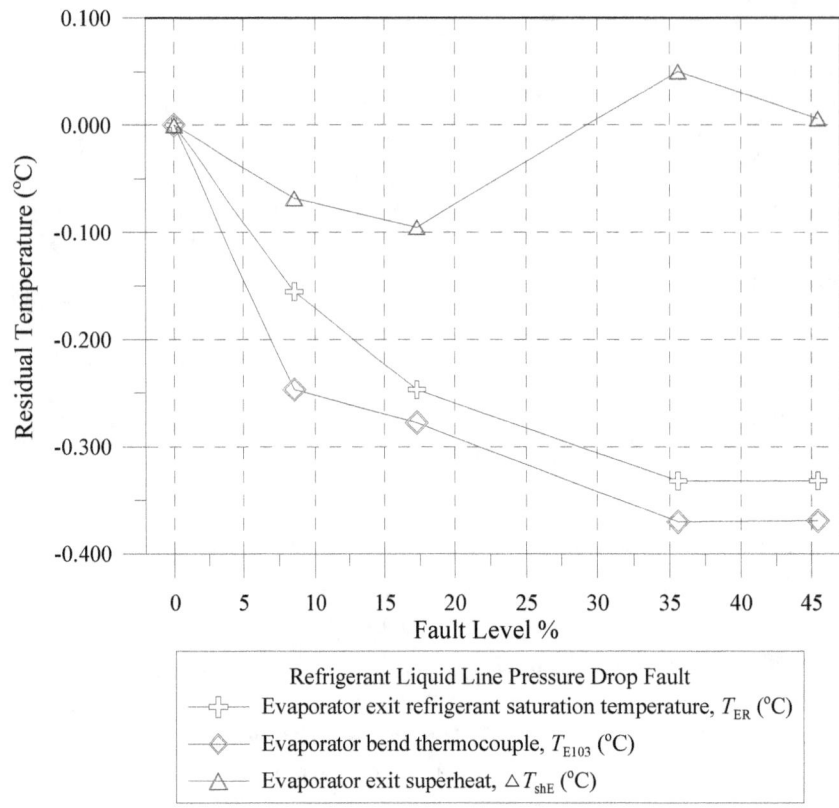

Figure 5.4.2. Residual of selected features with liquid line restriction faults at an indoor dry-bulb temperature of 21.1 °C and outdoor conditions of -8.3 °C/Dry: R[T_{ER}], R[T_{E103}], and R[ΔT_{shE}]

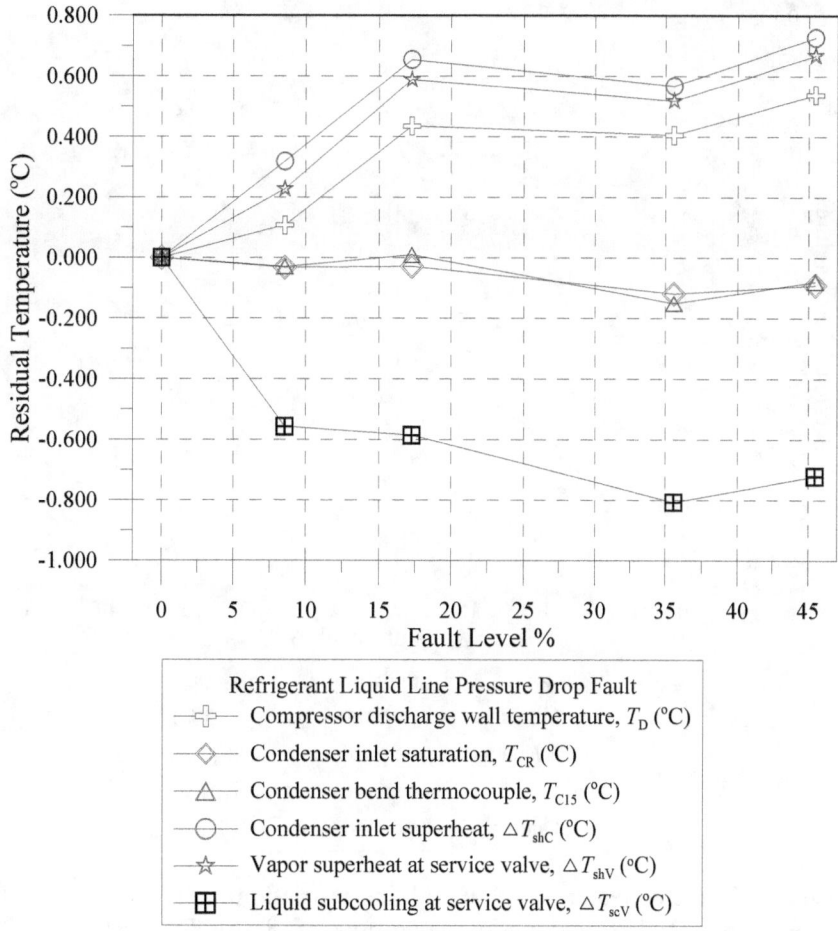

Figure 5.4.3. Residual of selected features with liquid line restriction faults at an indoor dry-bulb temperature of 21.1 °C and outdoor conditions of -8.3 °C/Dry: R[T_D], R[T_{CR}], R[T_{C15}], R[ΔT_{shC}], R[ΔT_{shV}], and R[ΔT_{scV}]

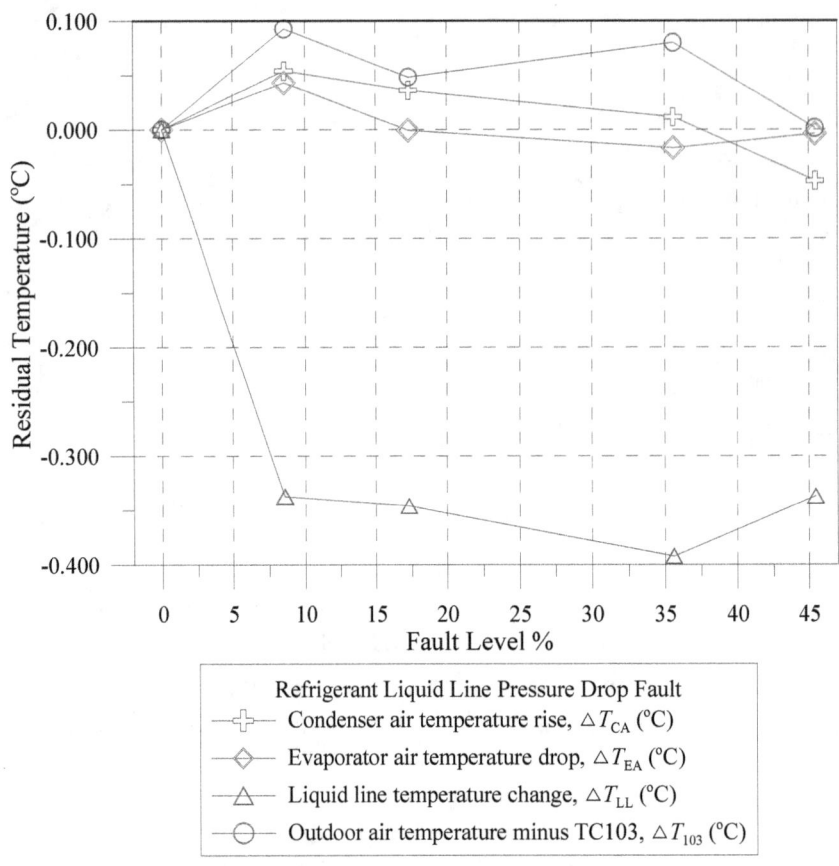

Figure 5.4.4. Residual of selected features with liquid line restriction faults at an indoor dry-bulb temperature of 21.1 °C and outdoor conditions of -8.3 °C/Dry: R[ΔT_{CA}], R[ΔT_{EA}], R[ΔT_{LL}], and R[ΔT_{103}]

114

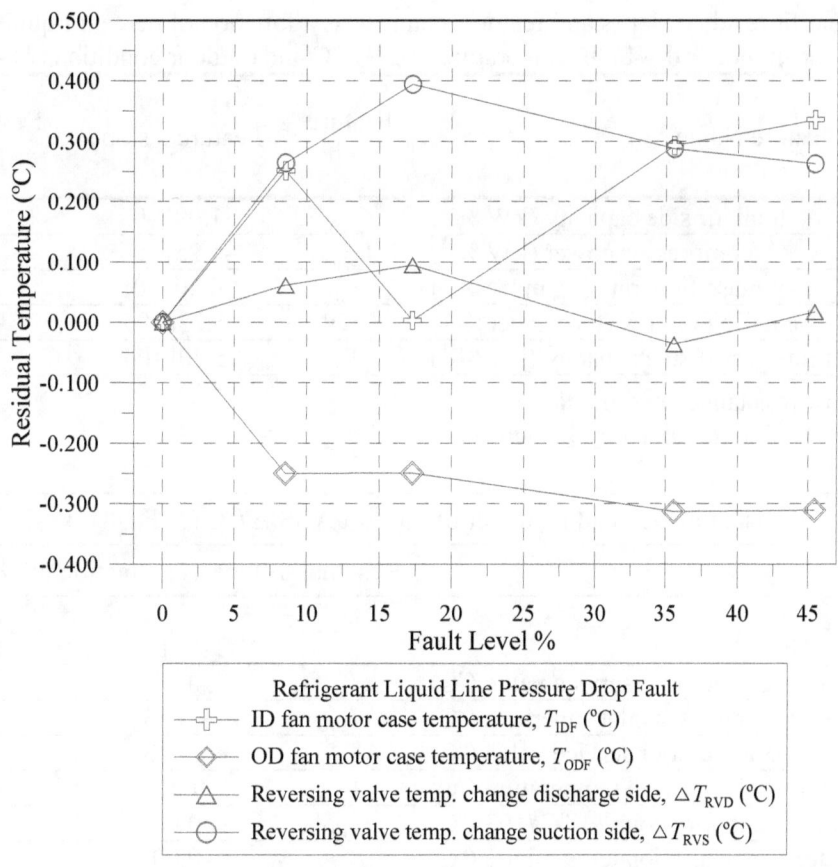

Figure 5.4.5. Residual of selected features with liquid line restriction faults at an indoor dry-bulb temperature of 21.1 °C and outdoor conditions of -8.3 °C/Dry: $R[T_{IDF}]$, $R[T_{ODF}]$, $R[\Delta T_{RVD}]$, and $R[\Delta T_{RVS}]$

Table 5.4.1. Linear fit residual slopes and features changes as a function of percent liquid line restriction fault at an indoor dry-bulb temperature of 21.1 °C and outdoor conditions of -8.3 °C/Dry

Feature Name	Feature Symbol	Feature's slope	Feature's % Change @ Max Fault Level
total air side capacity (kW %$^{-1}$)	Q_{CA}	-1.09E-03	-1.1
compressor power (kW %$^{-1}$)	W_{comp}	-1.25E-05	0.0
refrigerant mass flow rate (kg min^{-1} %$^{-1}$)	m_R	-6.40E-05	-0.2
coefficient of performance(%-1)	COP	5.22E-05	0.1
indoor unit refrigerant side capacity (kW %$^{-1}$)	Q_{CR}	5.70E-05	0.0

Valve used to vary flow resistance near middle of refrigerant liquid line. Fault determined by % increase in refrigerant pressure drop.

Max Fault Level: 45.4 %

Listed in Descending Order of Largest ABS(ΔT°C (% Fault)$^{-1}$)

		Residual's slope as a function of % fault level	
		Δ°C (% Fault)$^{-1}$	Feature % Change @ Max Fault Level
liquid line subcooling at outdoor service valve (°C)	ΔT_{scV}	-0.014	-11.1
condenser inlet superheat (°C)	ΔT_{shC}	0.013	2.9
vapor superheat at outdoor service valve (°C)	ΔT_{shV}	0.013	2.2
compressor discharge wall temperature (°C)	T_D	0.011	0.7
evaporator bend thermocouple, TC#103 (°C)	T_{E103}	-0.007	-8.1
evaporator exit saturation temperature (°C)	T_{ER}	-0.007	-14.4
ID fan motor case temperature (°C)	T_{IDF}	0.006	0.4
liquid line temperature drop (°C)	ΔT_{LL}	-0.006	-8.0
OD fan motor case temperature (°C)	T_{ODF}	-0.005	-0.9
reversing valve temperature change, suction side (°C)	ΔT_{RVS}	0.004	12.6
condenser bend thermocouple, TC#15 (°C)	T_{C15}	-0.003	-0.2
condenser inlet saturation temperature (°C)	T_{CR}	-0.002	-0.2
evaporator exit superheat (°C)	ΔT_{shE}	0.001	0.1
condenser air temperature rise (°C)	ΔT_{CA}	-0.001	-0.5
reversing valve temperature change, discharge side (°C)	ΔT_{RVD}	-0.001	1.0
evaporator air temperature drop (°C)	ΔT_{EA}	-0.001	-0.1
outdoor temperature minus TC#103 (°C)	ΔT_{103}	0.000	0.0

5.4.2 Indoor Dry-Bulb Temperature of 21.1 °C at Outdoor Conditions of 8.3 °C/72.5 % RH

Figure 5.4.6 shows the change in heating capacity, refrigerant-side heating capacity, compressor power, COP, and refrigerant mass flow rate as a function of the percent increase in refrigerant liquid line pressure drop. The corrective action taken by the thermostatic expansion valve (TXV) is shown by the oscillations in refrigerant mass flow rate and resulting oscillations of other features as the liquid line pressure drop was increased. Refrigerant-side heating capacity and COP changed by only -1.2 % and -1.8 %, respectively, with a 48.4 % increase in refrigerant pressure drop. A large fault level was indicated even though the absolute change in liquid line pressure drop was less than 40 kPa. Suction and discharge pressure absolute values changed by less than 38 kPa at the highest fault level.

Figure 5.4.7 shows the residuals of T_{ER}, T_{E103}, and ΔT_{shE} as a function of liquid line refrigerant pressure drop. The evaporator exit superheat, ΔT_{shE}, oscillated within 0.2 °C to -0.6 °C of its NFSS value due to corrective actions by the thermostatic expansion valve. The evaporator exit saturation temperature was insignificantly affected by this fault.

Figures 5.4.7 through 5.4.10 show residuals of selected features because of their significant oscillations due to the corrective action by the TXV, these features are not suitable for application in an FDD scheme.

Table 5.4.2 shows the residual's linear slopes and the percent changes in the system characteristics and temperatures for the liquid line pressure drop fault at outdoor conditions of 8.3 °C/72.5 % RH. The greatest change in residual (greatest slope) occurred for the condenser inlet saturation temperature; its value was -0.15 °C %⁻¹. Several features had large changes in value at the 48 % maximum fault level; T_{ER}, ΔT_{RVS}, ΔT_{CA}, and ΔT_{scV} changed by more than 10 % from their NFSS values, but the change in their residual values with respect to increasing fault level was small (less than 0.02 °C %⁻¹). All of the figures show that the liquid line refrigerant flow restriction fault has a minimal effect on the system. The high fault level is not indicative of the small absolute change in liquid line pressure drop; low mass flow rates and low ambient conditions produce low pressure drop in the liquid line and thus even a small absolute change in pressure drop will produce a large percentage change.

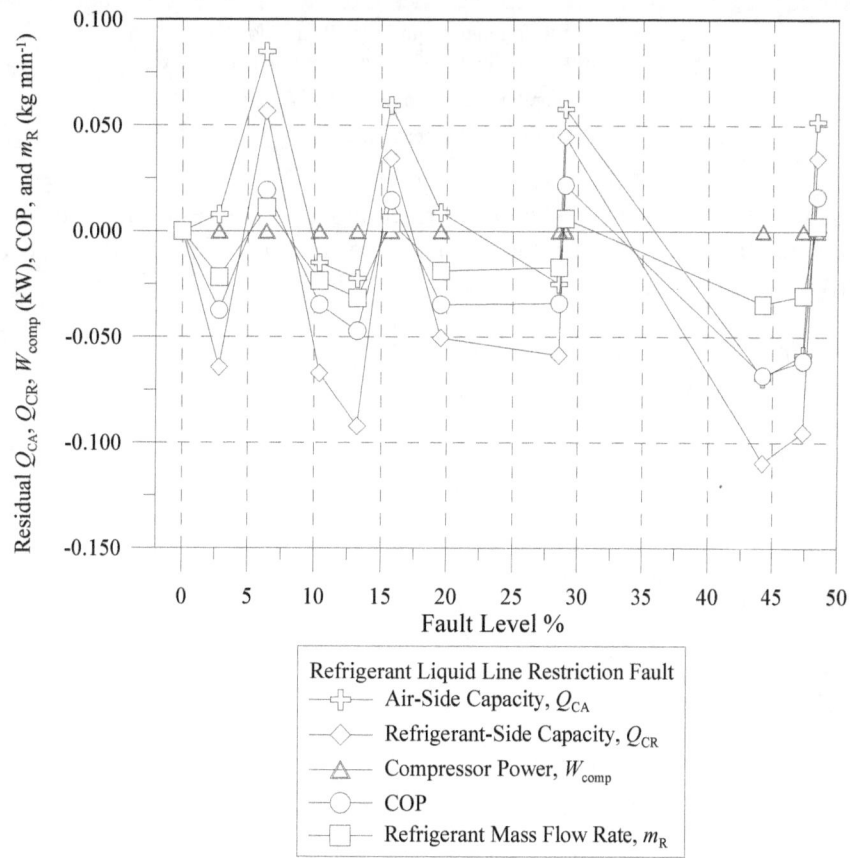

Figure 5.4.6. Residual of selected features with refrigerant liquid line restriction faults imposed at an indoor dry-bulb temperature of 21.1 °C and outdoor conditions of 8.3 °C/72.5 % RH: R[Q_{CA}], R[Q_{CR}], R[W_{comp}], R[COP], and R[m_R]

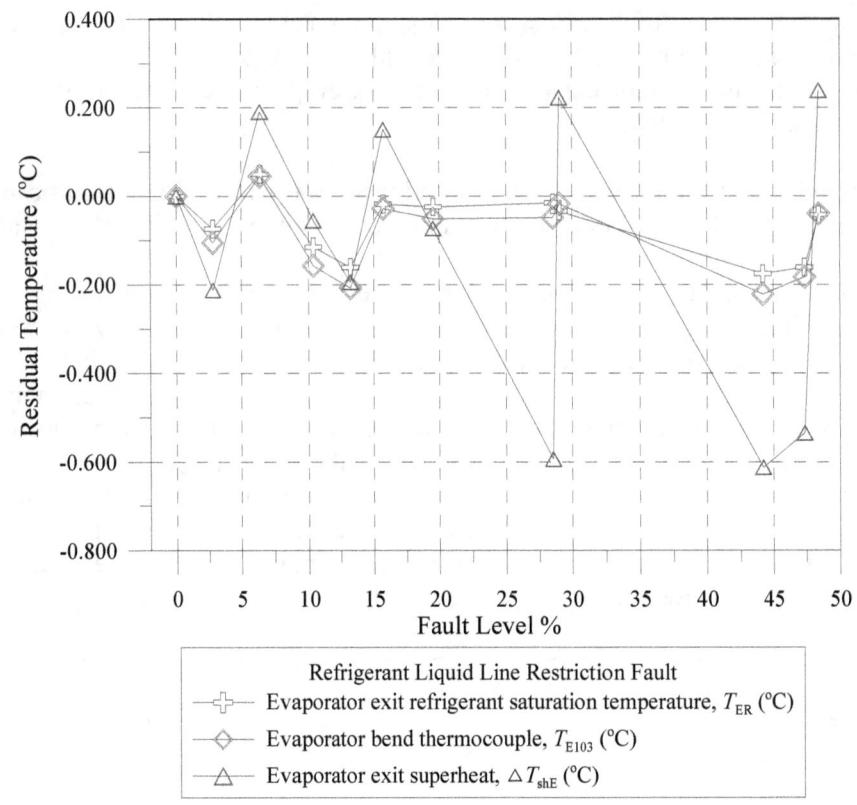

Figure 5.4.7. Residual of selected features with a refrigerant liquid line restriction fault at indoor dry-bulb temperature of 21.1 °C and outdoor conditions of 8.3 °C/72.5 % RH: R[T_{ER}], R[T_{E103}], and R[ΔT_{shE}]

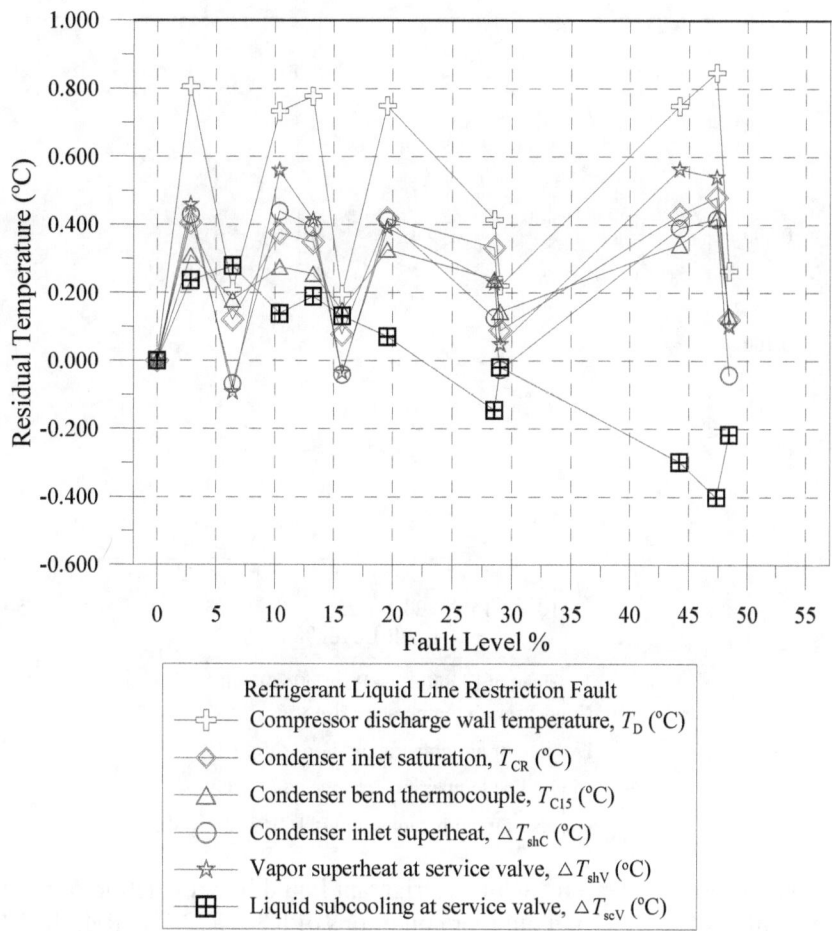

Figure 5.4.8. Residual of selected features with a refrigerant liquid line restriction fault at indoor dry-bulb temperature of 21.1 °C and outdoor conditions of 8.3 °C/72.5 % RH: R[T_D], R[T_{CR}], R[T_{C15}], R[ΔT_{shC}], R[ΔT_{shV}], and R[ΔT_{scV}]

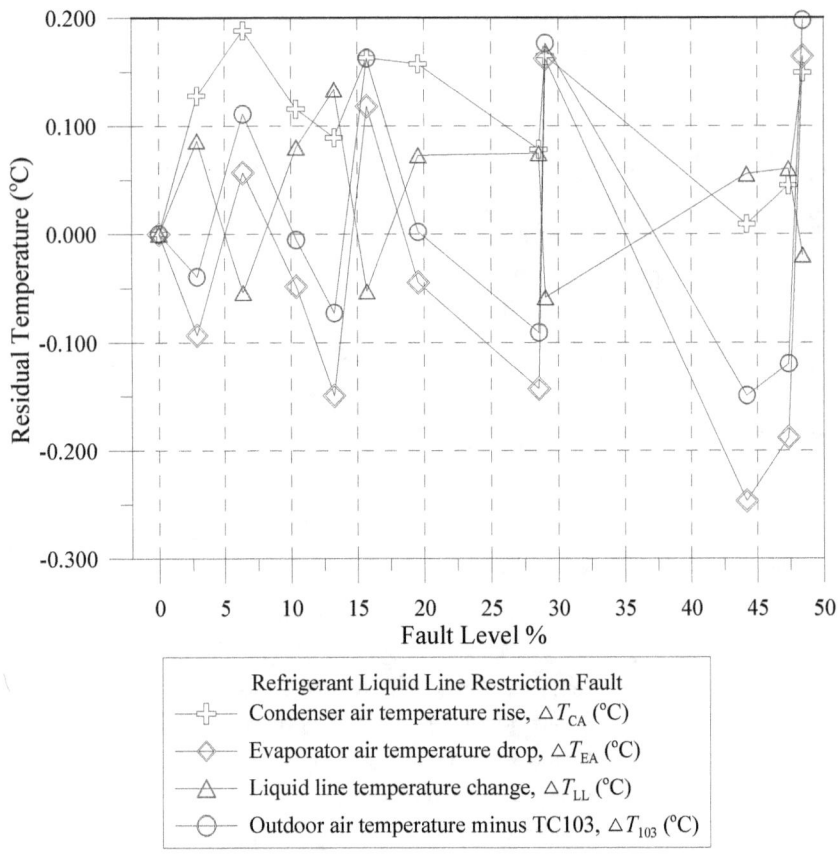

Figure 5.4.9. Residual of selected features with a refrigerant liquid line restriction fault at indoor dry-bulb temperature of 21.1 °C and outdoor conditions of 8.3 °C/72.5 % RH: R[ΔT_{CA}], R[ΔT_{EA}], R[ΔT_{LL}], and R[ΔT_{103}]

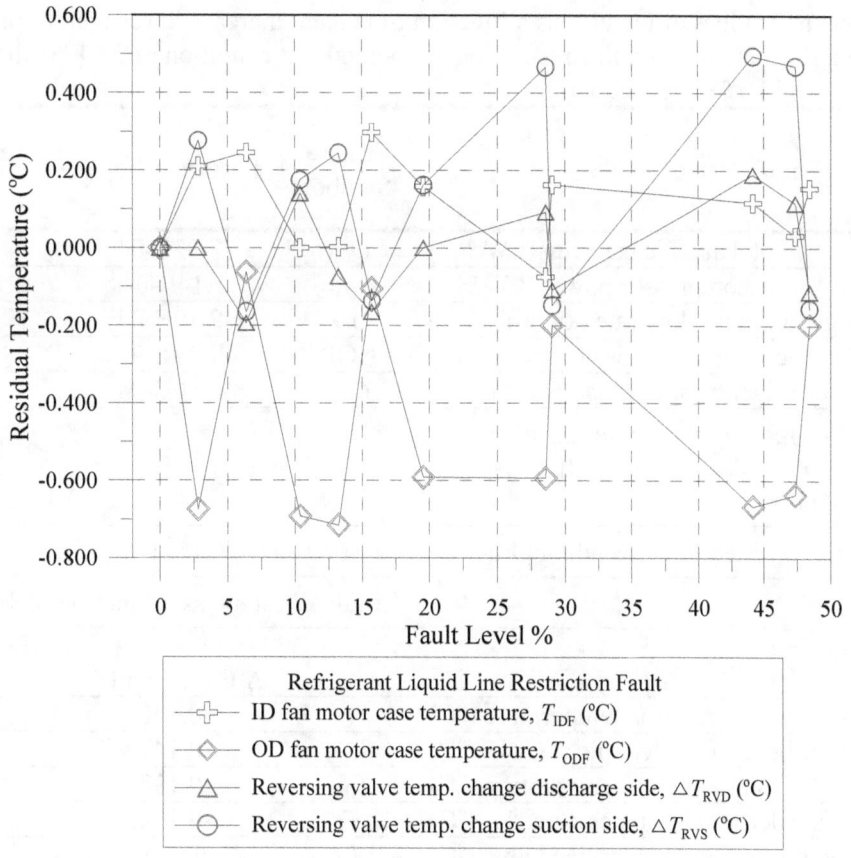

Figure 5.4.10. Residual of selected features with a refrigerant liquid line restriction fault at indoor dry-bulb temperature of 21.1 °C and outdoor conditions of 8.3 °C/72.5 % RH: R[T_{IDF}], R[T_{ODF}], R[ΔT_{RVD}], and R[ΔT_{RVS}]

Table 5.4.2. Linear fit slopes of features as a function of percent increase in refrigerant pressure drop due to a refrigerant liquid line restriction at nominal test condition of 21.1 °C indoor and 8.3 °C/72.5 % RH outdoor

Feature Name	Feature Symbol	Feature's slope	Feature % Change @ Max Fault Level
total air side capacity (kW %$^{-1}$)	Q_{CA}	-8.92E-04	-0.7
compressor power (kW %$^{-1}$)	W_{comp}	1.14E-04	1.1
refrigerant mass flow rate (kg min^{-1} %$^{-1}$)	m_R	-2.19E-04	-1.2
coefficient of performance(%$^{-1}$)	COP	-4.42E-04	-1.8
indoor unit refrigerant side capacity (kW %$^{-1}$)	Q_{CR}	-6.50E-04	-1.2

Valve used to vary flow resistance near middle of refrigerant liquid line. Fault determined by % increase in refrigerant pressure drop.

Max Fault Level: 48.4 %

Listed in Descending Order of Largest ABS(ΔT°C (% Fault)$^{-1}$)

		Residual's slope as a function of % fault level	
		Δ°C (% Fault)$^{-1}$	Feature % Change @ Max Fault Level
condenser inlet saturation temperature (°C)	T_{CR}	-0.148	-3.9
condenser bend thermocouple, TC#15 (°C)	T_{C15}	-0.147	-4.0
ID fan motor case temperature (°C)	T_{IDF}	-0.132	-2.3
condenser air temperature rise (°C)	ΔT_{CA}	-0.124	-10.9
vapor superheat at outdoor service valve (°C)	ΔT_{shV}	0.075	1.9
liquid line subcooling at outdoor service valve (°C)	ΔT_{scV}	0.063	-31.8
evaporator exit saturation temperature (°C)	T_{ER}	0.053	3.2
condenser inlet superheat (°C)	ΔT_{shC}	0.043	1.0
compressor discharge wall temperature (°C)	T_D	-0.039	-2.1
reversing valve temperature change, discharge side (°C)	ΔT_{RVD}	-0.036	-4.4
evaporator bend thermocouple, TC#103 (°C)	T_{E103}	0.028	1.4
outdoor temperature minus TC#103 (°C)	ΔT_{103}	-0.024	-5.6
evaporator air temperature drop (°C)	ΔT_{EA}	-0.021	-12.1
evaporator exit superheat (°C)	ΔT_{shE}	-0.018	-6.7
liquid line temperature drop (°C)	ΔT_{LL}	-0.006	8.5
OD fan motor case temperature (°C)	T_{ODF}	0.006	-0.1
reversing valve temperature change, suction side (°C)	ΔT_{RVS}	-0.004	16.1

5.5 Overcharged Refrigerant Fault (OC fault)

5.5.1 Indoor Dry-Bulb Temperature of 21.1 °C at Outdoor Conditions of -8.3 °C/Dry

The overcharged refrigerant fault was accomplished by setting the proper refrigerant charge in the cooling mode (according to manufacturer specs) and then weighing in more refrigerant to increase charge level during the heating mode. Figure 5.5.1 shows the change in air-side heating capacity, refrigerant-side heating capacity, compressor power, COP, and refrigerant mass flow rate as a function of the percent increase in refrigerant charge. Refrigerant-side heating capacity and COP changed by -0.1 % and -7.9 %, respectively, with a 30.4 % increase in refrigerant mass charged.

Figure 5.5.2 shows the residuals of T_{ER}, T_{E103}, and ΔT_{shE} as a function of refrigerant overcharge. The evaporator exit superheat, ΔT_{shE}, oscillates due to actions by the thermostatic expansion valve. Even though test conditions remained steady, the TXV may have tried to adjust superheat due to intermittent flooding of a refrigerant circuit. Even with proper charge the transition point from saturated to superheated refrigerant within a refrigerant circuit oscillates with respect to axial (streamwise) position. This has been observed with infrared photography of the evaporating refrigerant in indoor coils at steady air conditions. At the maximum fault level ΔT_{shE} was within 0.1 °C of its NFSS value. The pressure calculated value of the evaporator exit saturation temperature, T_{ER}, and the thermocouple measured value, T_{E103}, remained relatively constant with increasing fault level, and the residual value slopes were still less than 0.01 °C %$^{-1}$.

Figure 5.5.3 shows residuals for T_D, T_{CR}, T_{C15}, ΔT_{shC}, ΔT_{shV}, and ΔT_{scV}. This plot shows substantial changes in compressor discharge line wall temperatures with increasing fault level. Condenser inlet saturation temperature and superheat residuals also showed large changes. At the maximum fault level, the refrigerant subcooling at the service valve increased by more than 4.0 °C.

Figure 5.5.4 shows residuals for ΔT_{CA}, ΔT_{EA}, ΔT_{LL}, and ΔT_{103}. These features showed little change with increasing charge level.

Figure 5.5.5 shows residuals for T_{IDF}, T_{ODF}, ΔT_{RVD}, and ΔT_{RVS}. Compared to the NFSS value, the temperature change across the cold side of the four-way valve showed an immediate increase, and then decreased with increasing fault level. The mirroring of the suction side by the discharge side reversing valve temperature change faltered at the maximum fault level. The indoor fan motor case temperature also showed an increasing residual with increasing fault level, while the outdoor fan motor case temperature was decreasing.

Table 5.5.1 shows the residual's linear slopes and the percent changes in the system characteristics and temperatures for the overcharged refrigerant fault. Compressor discharge temperature and condenser inlet refrigerant saturation temperature were the features that changed the most; T_D and T_{CR} residual slopes were 0.264 °C %$^{-1}$ and 0.164 °C %$^{-1}$, respectively. Subcooling and superheat at the service valve along with superheat at the condenser inlet showed substantial change with their residual slopes being 0.133 °C %$^{-1}$, 0.089 °C %$^{-1}$, and 0.071 °C %$^{-1}$, respectively.

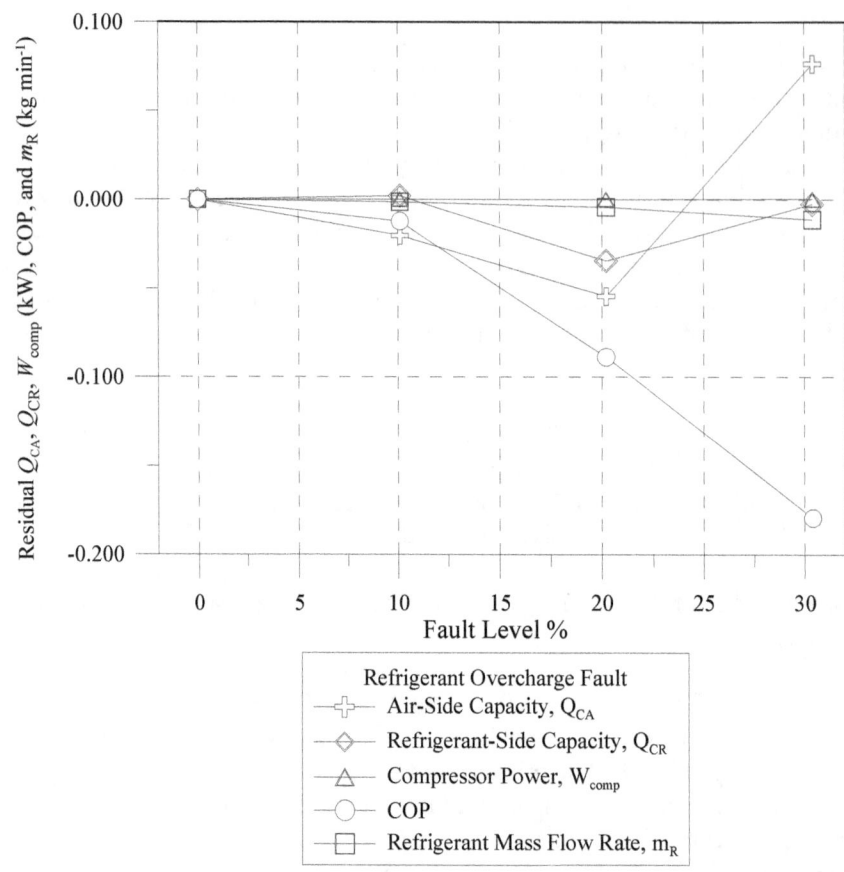

Figure 5.5.1. Residual of selected features with overcharged refrigerant faults imposed at an indoor dry-bulb temperature of 21.1 °C and outdoor conditions of -8.3 °C/Dry: R[Q_{CA}], R[Q_{CR}], R[W_{comp}], R[COP], and R[m_R]

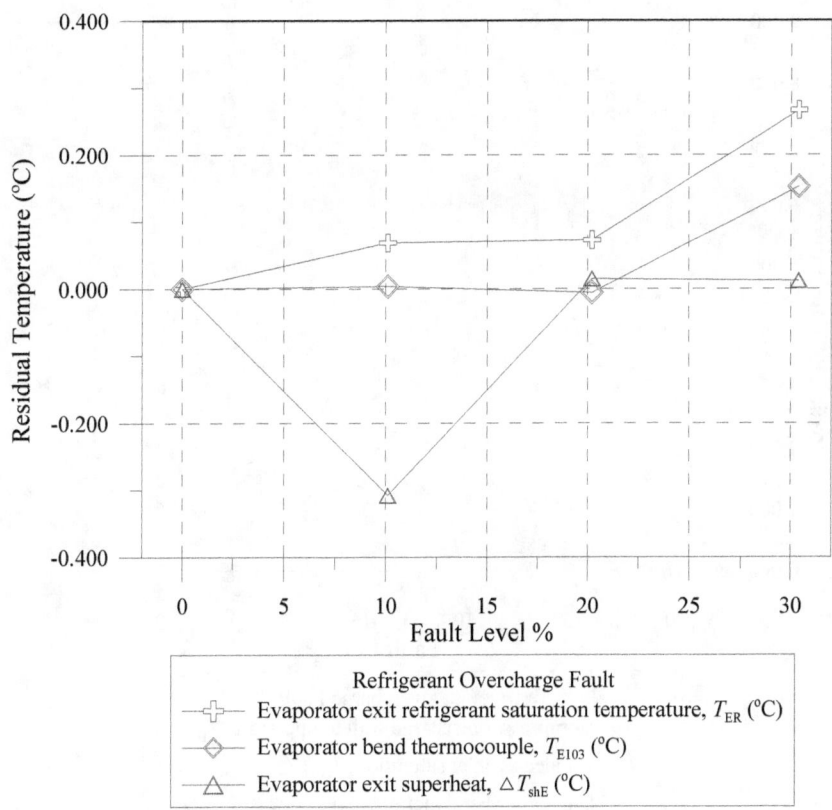

Figure 5.5.2. Residual of selected features with overcharged refrigerant faults at an indoor dry-bulb temperature of 21.1 °C and outdoor conditions of -8.3 °C/Dry: $R[T_{ER}]$, $R[T_{E103}]$, and $R[\Delta T_{shE}]$

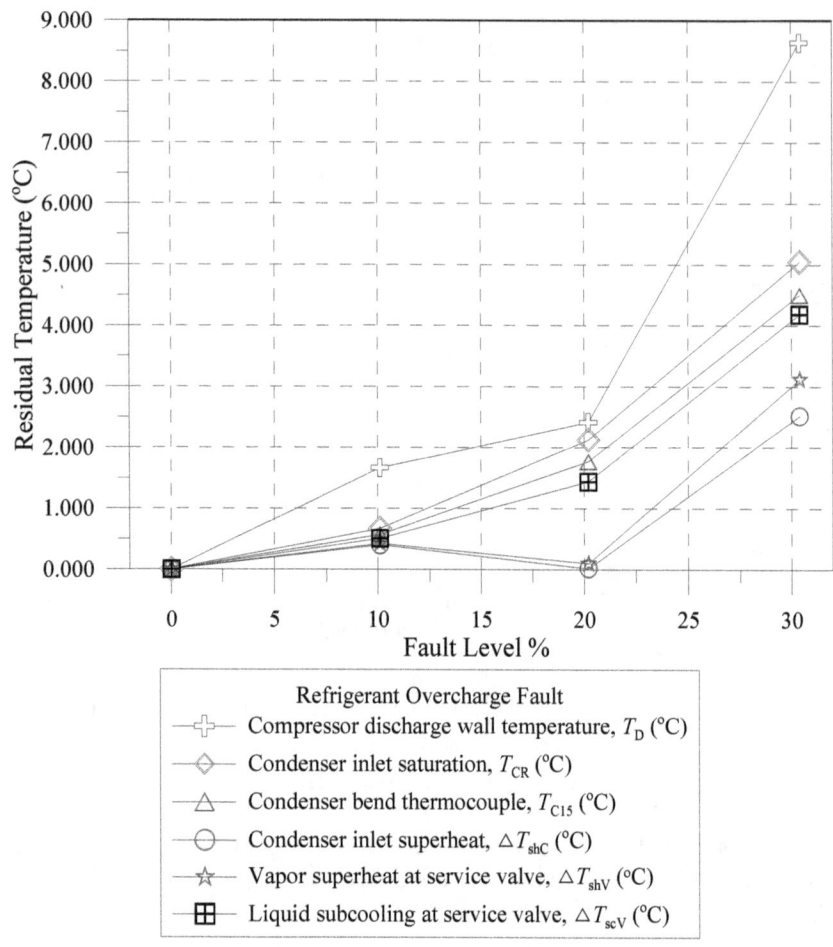

Figure 5.5.3. Residual of selected features with overcharged refrigerant faults at an indoor dry-bulb temperature of 21.1 °C and outdoor conditions of -8.3 °C/Dry: R[T_D], R[T_{CR}], R[T_{C15}], R[ΔT_{shC}], R[ΔT_{shV}], and R[ΔT_{scV}]

126

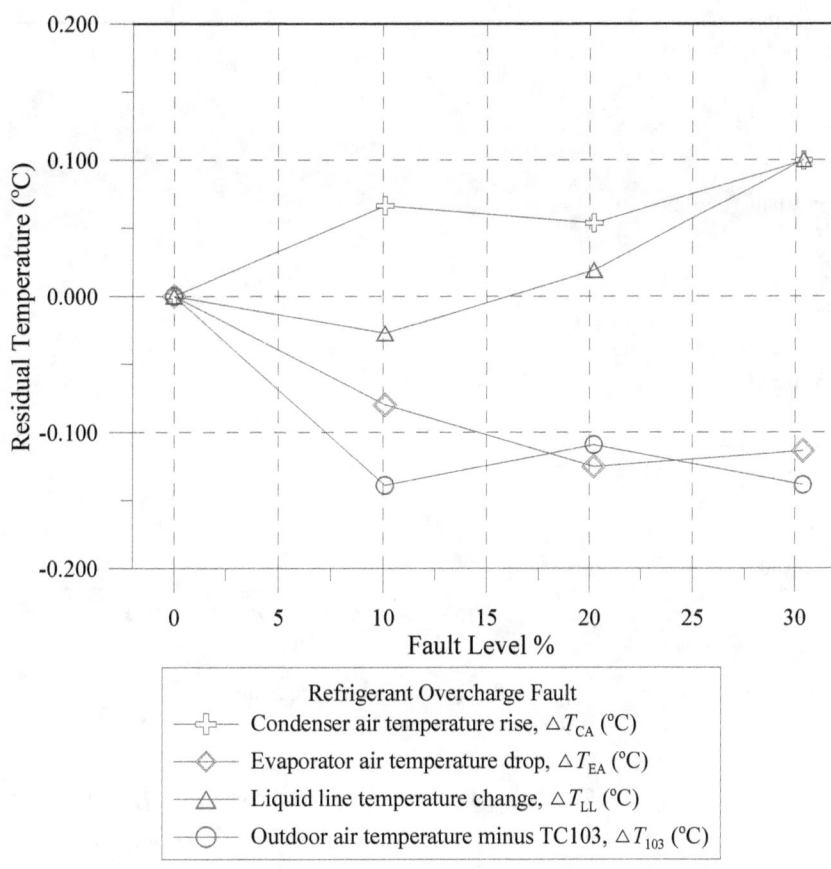

Figure 5.5.4. Residual of selected features with overcharged refrigerant faults at an indoor dry-bulb temperature of 21.1 °C and outdoor conditions of -8.3 °C/Dry: R[ΔT_{CA}], R[ΔT_{EA}], R[ΔT_{LL}], and R[ΔT_{103}]

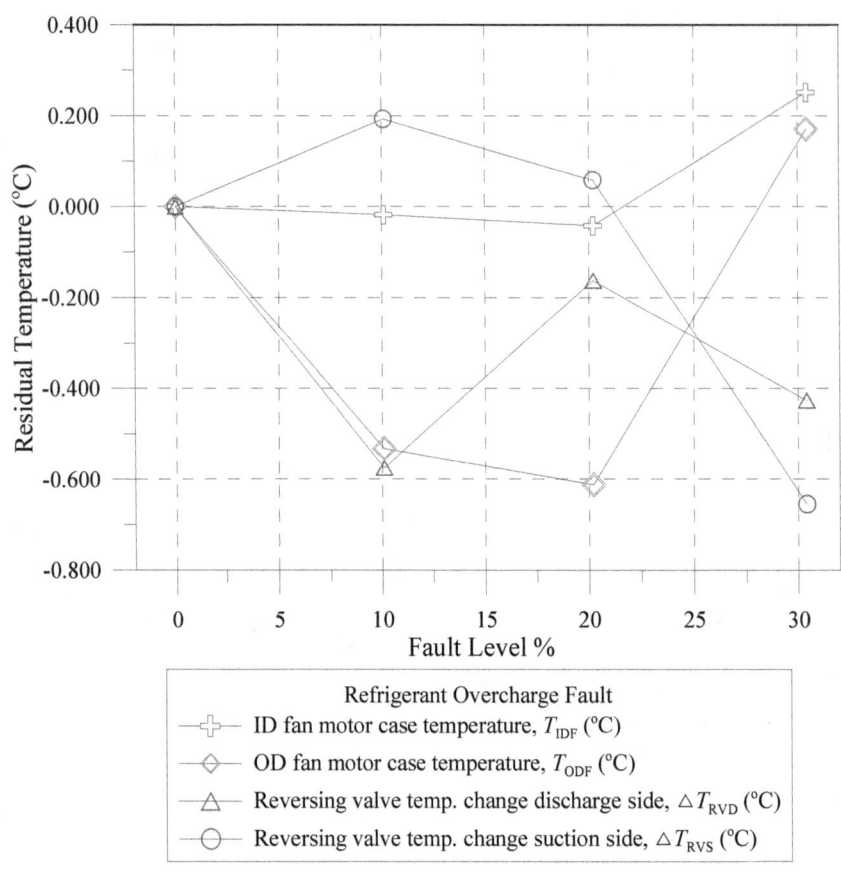

Figure 5.5.5. Residual of selected features with overcharged refrigerant faults at an indoor dry-bulb temperature of 21.1 °C and outdoor conditions of -8.3 °C/Dry: R[T_{IDF}], R[T_{ODF}], R[ΔT_{RVD}], and R[ΔT_{RVS}]

Table 5.5.1. Linear fit residual slopes and feature changes as a function of percent increase in refrigerant charge at an indoor dry-bulb temperature of 21.1 °C and outdoor conditions of -8.3 °C/Dry

Feature Name	Feature Symbol	Feature's slope	Feature's % Change @ Max Fault Level
total air side capacity (kW %$^{-1}$)	Q_{CA}	1.94E-03	1.5
compressor power (kW %$^{-1}$)	W_{comp}	6.15E-03	11.7
refrigerant mass flow rate (kg min^{-1} %$^{-1}$)	m_R	-3.60E-04	-0.8
coefficient of performance(%$^{-1}$)	COP	-6.07E-03	-7.9
indoor unit refrigerant side capacity (kW %$^{-1}$)	Q_{CR}	-4.32E-04	-0.1

Refrigerant mass was added. Fault determined by % increase above normal system charge level determined in the cooling mode.

Max Fault Level: 30.4 %

Listed in Descending Order of Largest ABS(ΔT°C (% Fault)$^{-1}$)

		Residual's slope as a function of % fault level	
		Δ°C (% Fault)$^{-1}$	Feature % Change @ Max Fault Level
compressor discharge wall temperature (°C)	T_D	0.264	11.2
condenser inlet saturation temperature (°C)	T_{CR}	0.164	10.2
condenser bend thermocouple, TC#15 (°C)	T_{C15}	0.145	9.1
liquid line subcooling at outdoor service valve (°C)	ΔT_{scV}	0.133	64.5
vapor superheat at outdoor service valve (°C)	ΔT_{shV}	0.089	10.4
condenser inlet superheat (°C)	ΔT_{shC}	0.071	10.0
reversing valve temperature change, suction side (°C)	ΔT_{RVS}	-0.021	-31.4
reversing valve temperature change, discharge side (°C)	ΔT_{RVD}	-0.009	-23.6
evaporator exit saturation temperature (°C)	T_{ER}	0.008	11.6
ID fan motor case temperature (°C)	T_{IDF}	0.007	0.3
evaporator bend thermocouple, TC#103 (°C)	T_{E103}	0.004	3.3
OD fan motor case temperature (°C)	T_{ODF}	0.004	0.5
outdoor temperature minus TC#103 (°C)	ΔT_{103}	-0.004	-2.9
evaporator air temperature drop (°C)	ΔT_{EA}	-0.004	-4.1
evaporator exit superheat (°C)	ΔT_{shE}	0.004	0.2
liquid line temperature drop (°C)	ΔT_{LL}	0.003	2.4
condenser air temperature rise (°C)	ΔT_{CA}	0.003	1.1

5.5.2   Indoor Dry-Bulb Temperature of 21.1 °C at Outdoor Conditions of 8.3 °C/Dry

Figure 5.5.6 shows the change in air-side heating capacity, refrigerant-side heating capacity, compressor power, COP, and refrigerant mass flow rate as a function of the percent increase in refrigerant charge within the system. Refrigerant-side heating capacity and COP changed by -0.1 % and -14.4 %, respectively, with a 30.4 % increase in refrigerant mass charged. COP decreased due to an increase in compressor power of 22 % at the maximum fault level.

Figure 5.5.7 shows the residuals of T_{ER}, T_{E103}, and ΔT_{shE} as a function of refrigerant overcharge. The evaporator exit superheat, ΔT_{shE}, increases then decreases due to corrective actions by the thermostatic expansion valve. At the maximum fault level ΔT_{shE} was within 0.1 °C of its NFSS value. The pressure calculated value of the evaporator exit saturation temperature, T_{ER}, and the thermocouple measured value, T_{E103}, remained relatively constant with increasing fault level, and the residual value slopes were still less than 0.01 C %$^{-1}$. The saturation temperature residual clearly shows a negative slope then a positive slope as fault level increases.

Figure 5.5.8 shows the residuals of T_D, T_{CR}, T_{C15}, ΔT_{shC}, ΔT_{shV}, and ΔT_{scV}. This plot shows there are substantial changes in compressor discharge line wall temperatures with increasing fault level. Condenser inlet saturation temperature, superheat and liquid line subcooling at the service valve residuals also showed large changes. At the maximum fault level, the refrigerant subcooling at the service valve increased by more than 7.0 °C.

Figure 5.5.9 shows the residuals of ΔT_{CA}, ΔT_{EA}, ΔT_{LL}, and ΔT_{103}. These features showed little change with increasing charge level relative to the changes that occurred with the compressor discharge temperature.

Figure 5.5.10 shows the residuals of T_{IDF}, T_{ODF}, ΔT_{RVD}, and ΔT_{RVS}. Compared to the NFSS value, the temperature change across the cold side of the four-way valve decreased with increasing fault level. The mirroring of the suction side by the discharge side reversing valve temperature change held at all fault levels. Overall, the residuals are small, within 1°C of the minimum level of fault.

Table 5.5.2 shows the residual's linear slopes and the percent changes in the system characteristics and temperatures for the overcharged refrigerant fault. Compressor discharge temperature and condenser inlet refrigerant saturation temperature were the features that changed the most; T_D and T_{CR} residual slopes were 0.403 C %$^{-1}$ and 0.270 C %$^{-1}$, respectively. Subcooling and superheat at the service valve along with superheat at the condenser inlet showed substantial change with their residual slopes being 0.229 C %$^{-1}$, 0.160 C %$^{-1}$, and 0.144 C %$^{-1}$, respectively.

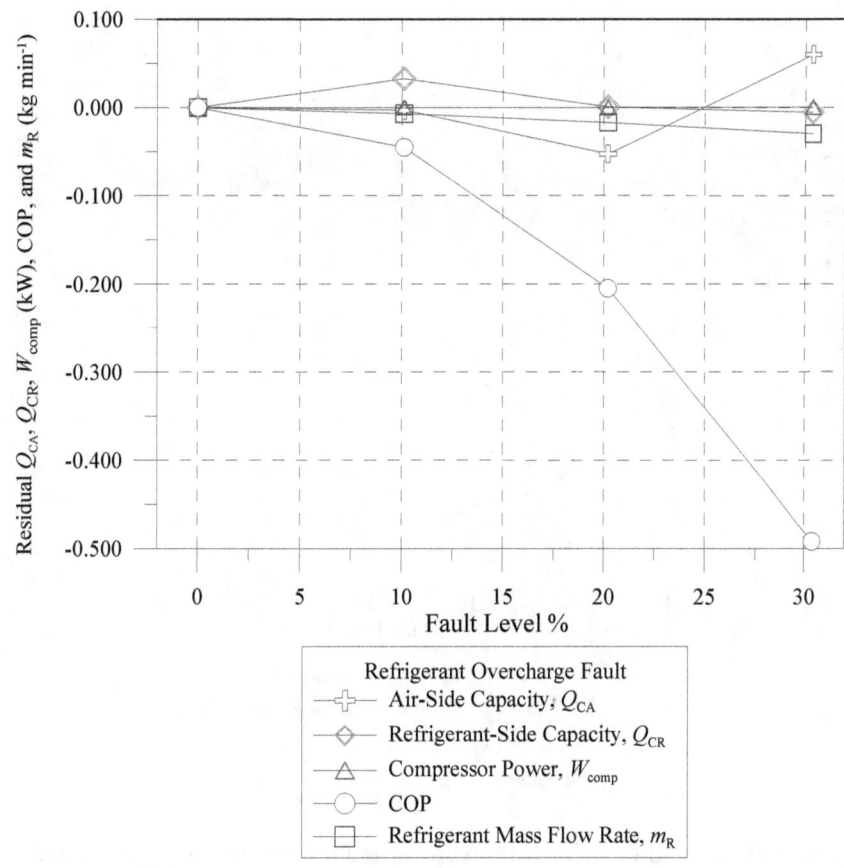

Figure 5.5.6. Residual of selected features with overcharged refrigerant faults imposed at an indoor dry-bulb temperature of 21.1 °C and outdoor conditions of 8.3 °C/Dry: R[Q_{CA}], R[Q_{CR}], R[W_{comp}], R[COP], and R[m_R]

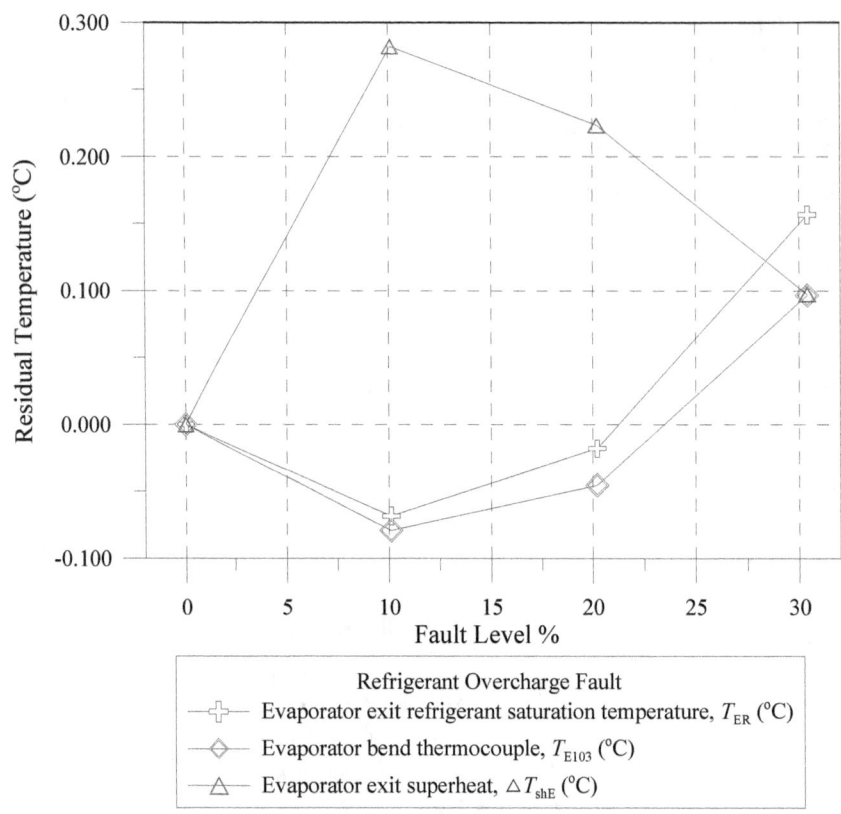

Figure 5.5.7. Residual of selected features with overcharged refrigerant faults at indoor dry-bulb temperature of 21.1 °C and outdoor conditions of 8.3 °C/Dry: R[T_{ER}], R[T_{E103}], and R[ΔT_{shE}]

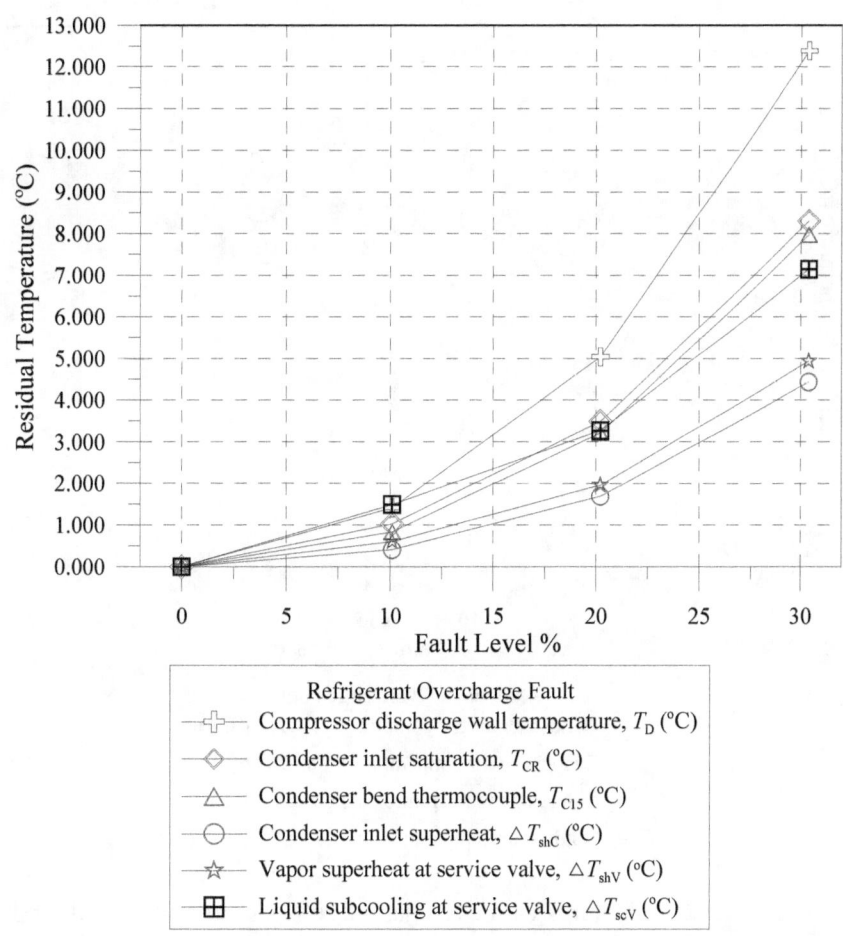

Figure 5.5.8. Residual of selected features with overcharged refrigerant faults at indoor dry-bulb temperature of 21.1 °C and outdoor conditions of 8.3 °C/Dry: R[T_D], R[T_{CR}], R[T_{C15}], R[ΔT_{shC}], R[ΔT_{shV}], and R[ΔT_{scV}]

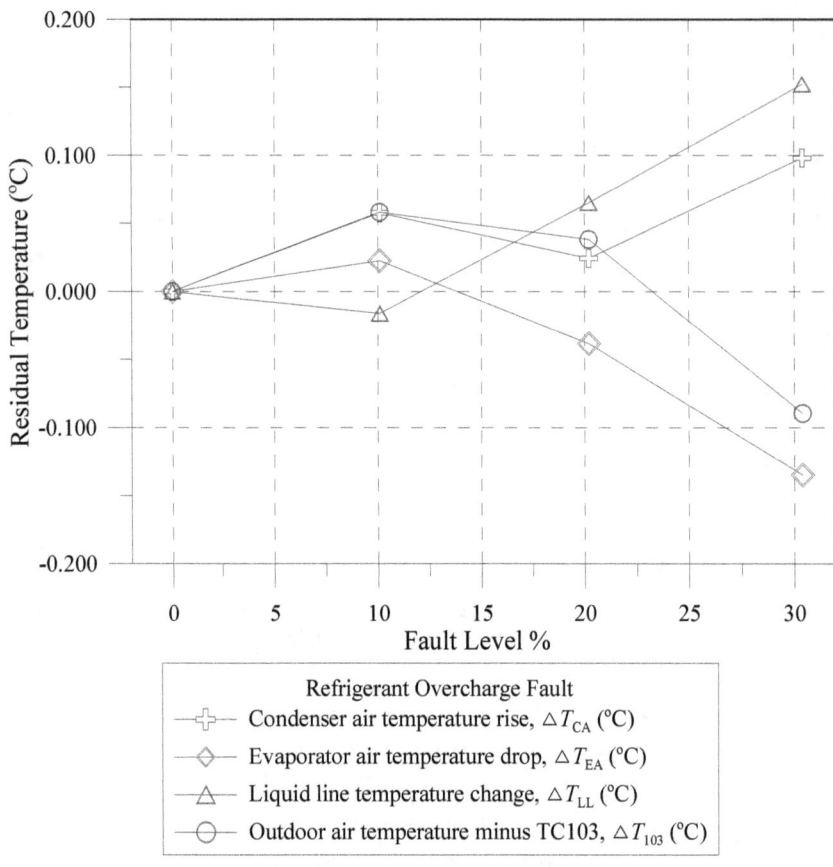

Figure 5.5.9. Residual of selected features with overcharged refrigerant faults at indoor dry-bulb temperature of 21.1 °C and outdoor conditions of 8.3 °C/Dry: R[ΔT_{CA}], R[ΔT_{EA}], R[ΔT_{LL}], and R[ΔT_{103}]

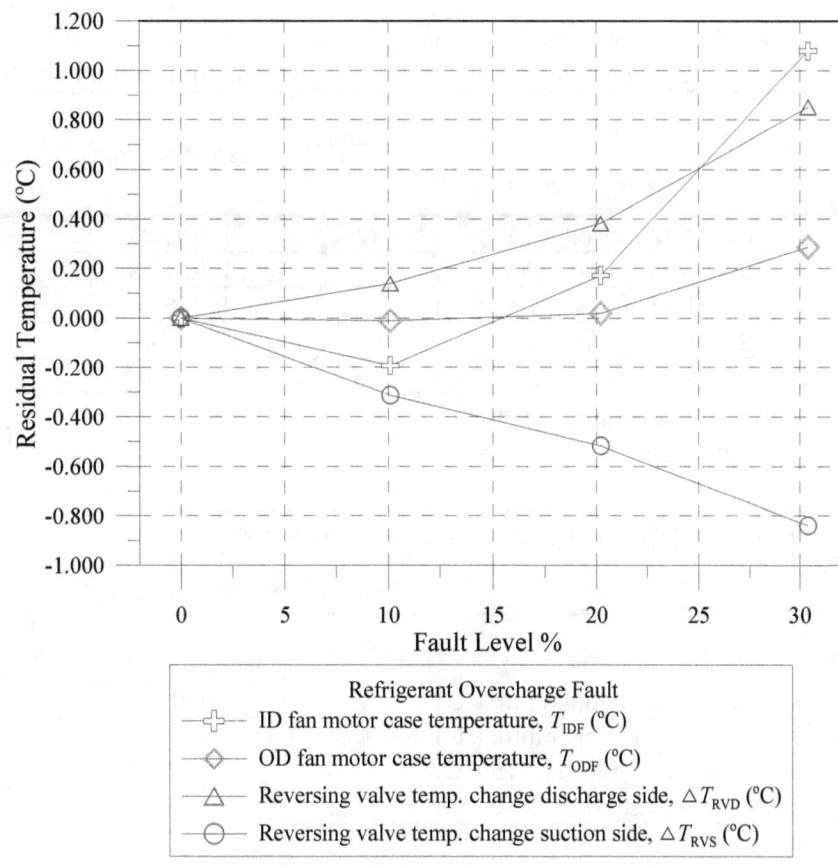

Figure 5.5.10. Residual of selected features with overcharged refrigerant faults at indoor dry-bulb temperature of 21.1 °C and outdoor conditions of 8.3 °C/Dry: R[T_{IDF}], R[T_{ODF}], R[ΔT_{RVD}], and R[ΔT_{RVS}]

Table 5.5.2. Linear fit residual slopes and feature changes as a function of percent increase in refrigerant charge at indoor dry-bulb temperature of 21.1 °C and outdoor conditions of 8.3 °C/Dry

Feature Name	Feature Symbol	Feature's slope	Feature's % Change @ Max Fault Level
total air side capacity (kW %$^{-1}$)	Q_{CA}	1.30E-03	0.7
compressor power (kW %$^{-1}$)	W_{comp}	1.30E-02	22.1
refrigerant mass flow rate (kg min^{-1} %$^{-1}$)	m_R	-9.83E-04	-1.3
coefficient of performance(%$^{-1}$)	COP	-1.62E-02	-14.5
indoor unit refrigerant side capacity (kW %$^{-1}$)	Q_{CR}	-4.85E-04	-0.1

Refrigerant mass was added. Fault determined by % increase above normal system charge level determined in the cooling mode.	Max Fault Level: +30.4 %

Listed in Descending Order of Largest ABS(ΔT°C (% Fault)$^{-1}$)

		Residual's slope as a function of % fault level	
		Δ°C (% Fault)$^{-1}$	Feature % Change @ Max Fault Level
compressor discharge wall temperature (°C)	T_D	0.403	15.8
condenser inlet saturation temperature (°C)	T_{CR}	0.270	14.8
condenser bend thermocouple, TC#15 (°C)	T_{C15}	0.260	14.0
liquid line subcooling at outdoor service valve (°C)	ΔT_{scV}	0.229	175.6
vapor superheat at outdoor service valve (°C)	ΔT_{shV}	0.160	19.0
condenser inlet superheat (°C)	ΔT_{shC}	0.144	18.6
ID fan motor case temperature (°C)	T_{IDF}	0.036	1.3
reversing valve temperature change, discharge side (°C)	ΔT_{RVD}	0.028	26.2
reversing valve temperature change, suction side (°C)	ΔT_{RVS}	-0.027	-112.1
OD fan motor case temperature (°C)	T_{ODF}	0.009	0.6
liquid line temperature drop (°C)	ΔT_{LL}	0.005	7.0
evaporator exit saturation temperature (°C)	T_{ER}	0.005	0.9
evaporator air temperature drop (°C)	ΔT_{EA}	-0.005	-2.7
evaporator bend thermocouple, TC#103 (°C)	T_{E103}	0.003	0.5
outdoor temperature minus TC#103 (°C)	ΔT_{103}	-0.003	-1.3
condenser air temperature rise (°C)	ΔT_{CA}	0.003	0.7
evaporator exit superheat (°C)	ΔT_{shE}	0.002	1.5

5.6 Undercharged Refrigerant Fault (UC fault)

5.6.1   Indoor Dry-Bulb Temperature of 21.1 °C at Outdoor Conditions of -8.3 °C/Dry

The undercharged refrigerant fault was accomplished by setting the proper refrigerant charge in the cooling mode (according to manufacturer specs) and then removing refrigerant. Figure 5.6.1 shows the change in air-side heating capacity, refrigerant-side heating capacity, compressor power, COP, and refrigerant mass flow rate as a function of the percent decrease in refrigerant charge (% decrease is a negative number). Air-side heating capacity and air-side COP changed by -4.9 % and -2.8 %, respectively, with a -30.3 % change in refrigerant mass charged. Normally, refrigerant and air side measurements would be within ±3 % or better, but refrigerant mass flow measurements, and thus refrigerant-side capacity, was affected by the presence of bubbles in the liquid line. These bubbles caused erroneous mass flow rate readings in the Coriolis meter, thus making any refrigerant-side measurements erroneous as well.

Figure 5.6.2 shows the residuals for T_{ER}, T_{E103}, and ΔT_{shE} as a function of refrigerant undercharge. The evaporator exit superheat, ΔT_{shE}, remains stable until a 20 % reduction in charge due to corrective action by the thermostatic expansion valve; increases in superheat may indicate that the TXV was at maximum opening and thus began operating like a fixed area expansion device. At the maximum fault level ΔT_{shE} increased to be 0.5 °C greater than its NFSS value. The pressure calculated value of the evaporator exit saturation temperature, T_{ER}, and the thermocouple measured value, T_{E103}, remained relatively constant with increasing fault level.

Figure 5.6.3 shows residuals for T_D, T_{CR}, T_{C15}, ΔT_{shC}, ΔT_{shV}, and ΔT_{scV}. Liquid line subcooling at the service valve, ΔT_{scV}, has the greatest change due to the undercharge fault. At the maximum fault level of -30.3 %, the liquid line subcooling drops almost 6 °C below its NFSS value. The residuals of the condenser inlet saturation temperature and its thermocouple counterpart, T_{CR} and T_{C15}, 0.045 °C %$^{-1}$ and 0.040 °C %$^{-1}$, respectively.

Figure 5.6.4 shows residuals for ΔT_{CA}, ΔT_{EA}, ΔT_{LL}, and ΔT_{103}. The figure shows that the liquid line temperature change is a good indicator for loss of refrigerant charge. This was the second largest slope seen for all features. It is interesting to note that very little variation in any feature occurs at up to a 10 % undercharge. The TXV is able to correct mass flow rate until, somewhere between 10 % and 20 % undercharge, the presence of two-phase refrigerant at the TXV inlet causes it to open fully and begin performing like a fixed area expansion device.

Figure 5.6.5 shows residuals for T_{IDF}, T_{ODF}, ΔT_{RVD}, and ΔT_{RVS}. The indoor fan case temperature residual shows some correlation to refrigerant undercharge. The only other feature here to show some reasonable linear change with undercharge is the cold side reversing valve temperature change, ΔT_{RVS}. This feature increases by more than 14 % from its NFSS value and has a residual slope of -0.009 °C %$^{-1}$.

Table 5.6.1 shows the residual's linear slopes and the percent changes in the system characteristics and temperatures for the undercharged refrigerant fault. The prominent features for the undercharge fault are liquid line subcooling, liquid line temperature drop from the outdoor service valve to the indoor TXV inlet, condenser inlet saturation temperature, and vapor line superheat at the service valve. Interestingly, the indoor fan case temperature residual was comparable to the service valve superheat residual for indicating an undercharge fault.

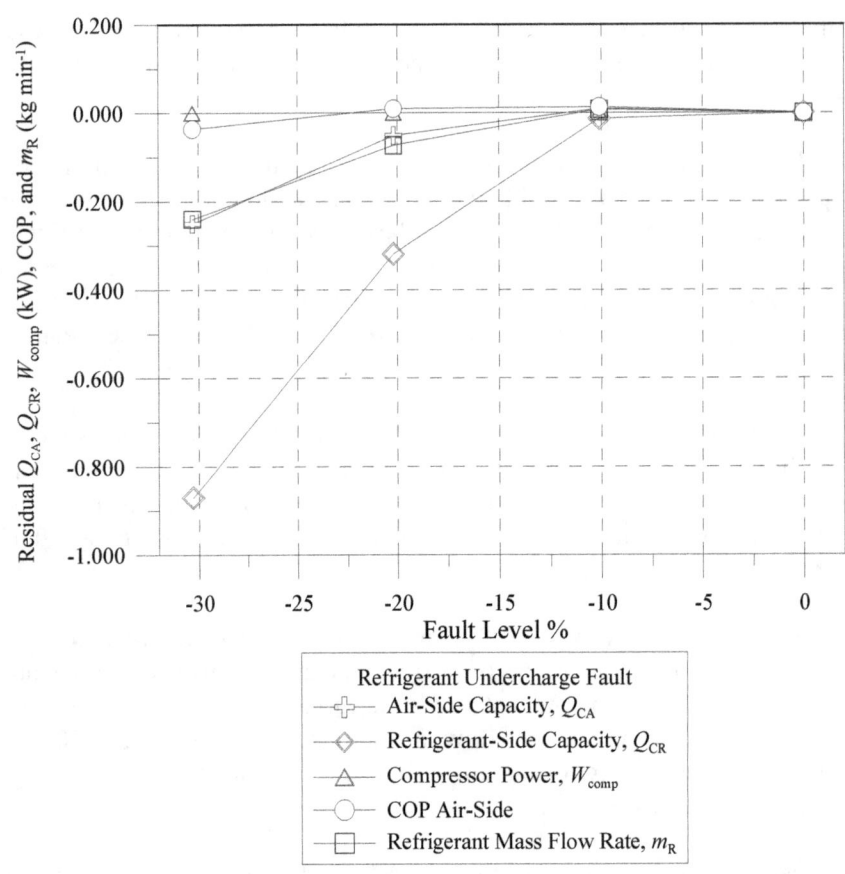

Figure 5.6.1. Residual of selected features with undercharged refrigerant faults imposed at an indoor dry-bulb temperature of 21.1 °C and outdoor conditions of -8.3 °C/Dry: R[Q_{CA}], R[Q_{CR}], R[W_{comp}], R[COP], and R[m_R]

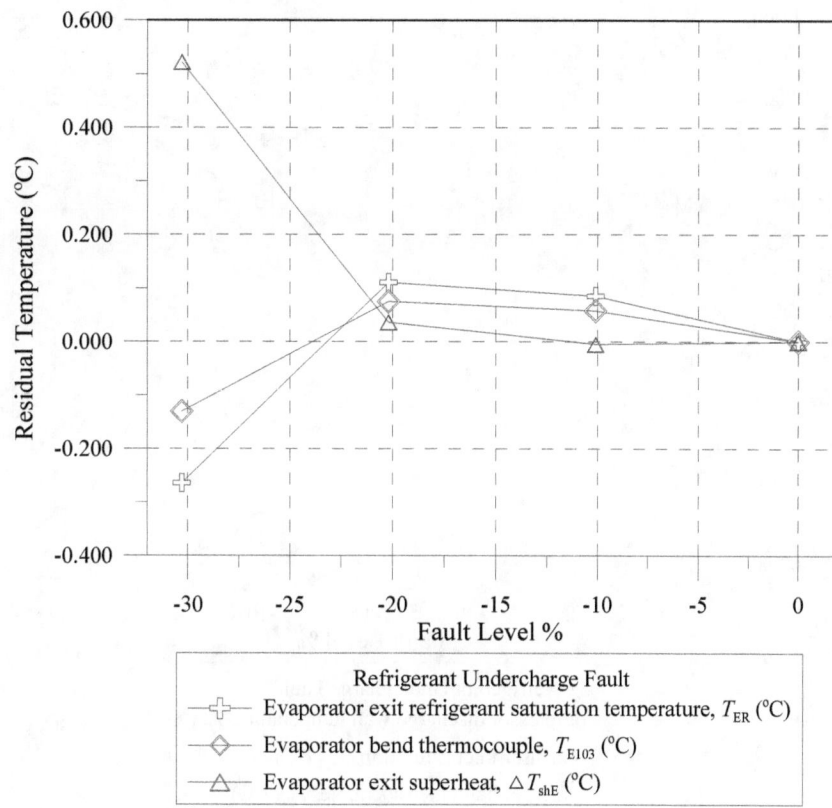

Figure 5.6.2. Residual of selected features with undercharged refrigerant faults at indoor dry-bulb temperature of 21.1 °C and outdoor conditions of -8.3 °C/Dry: R[T_{ER}], R[T_{E103}], and R[ΔT_{shE}]

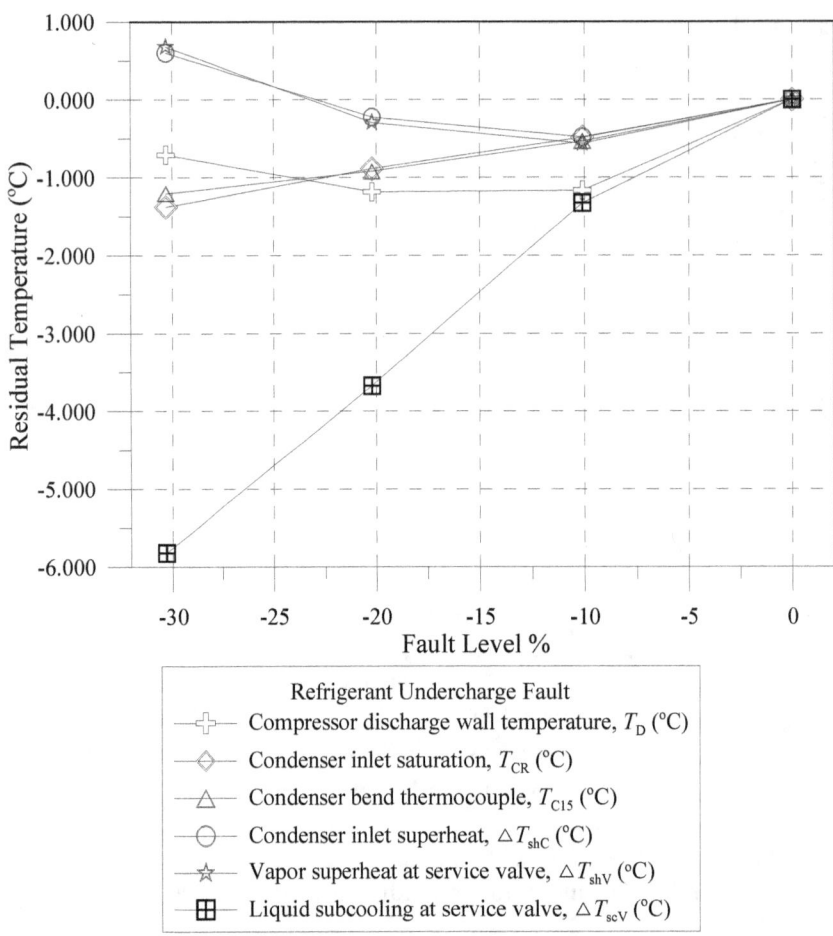

Figure 5.6.3. Residual of selected features with undercharged refrigerant faults at indoor dry-bulb temperature of 21.1 °C and outdoor conditions of -8.3 °C/Dry: R[T_D], R[T_{CR}], R[T_{C15}], R[ΔT_{shC}], R[ΔT_{shV}], and R[ΔT_{scV}]

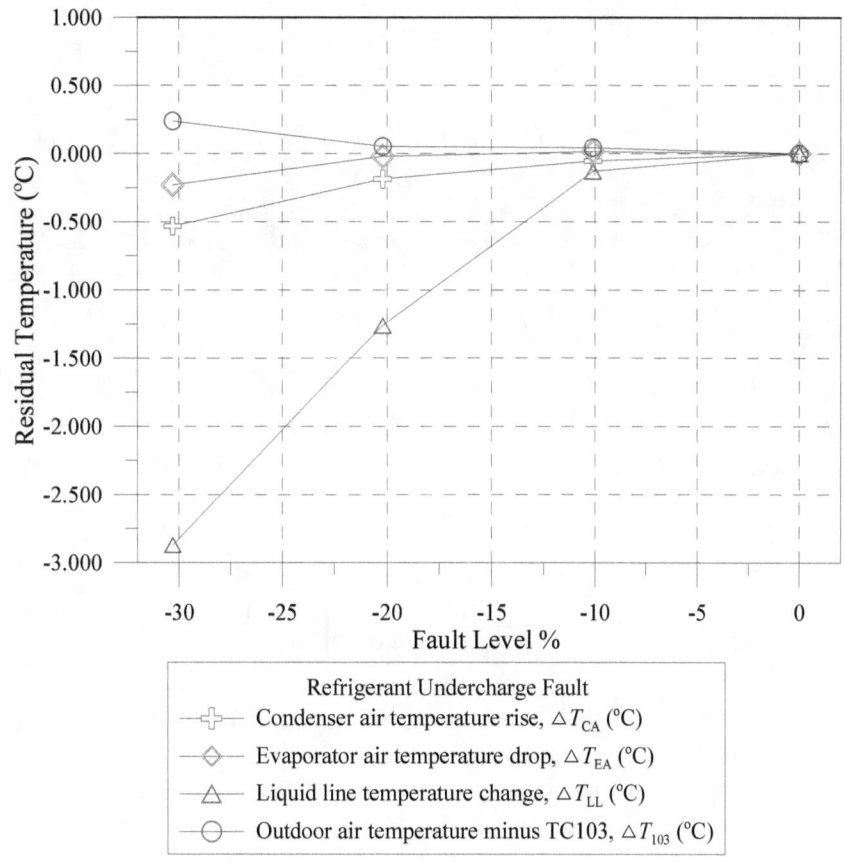

Figure 5.6.4. Residual of selected features with undercharged refrigerant faults at indoor dry-bulb temperature of 21.1 °C and outdoor conditions of -8.3 °C/Dry: R[ΔT_{CA}], R[ΔT_{EA}], R[ΔT_{LL}], and R[ΔT_{103}]

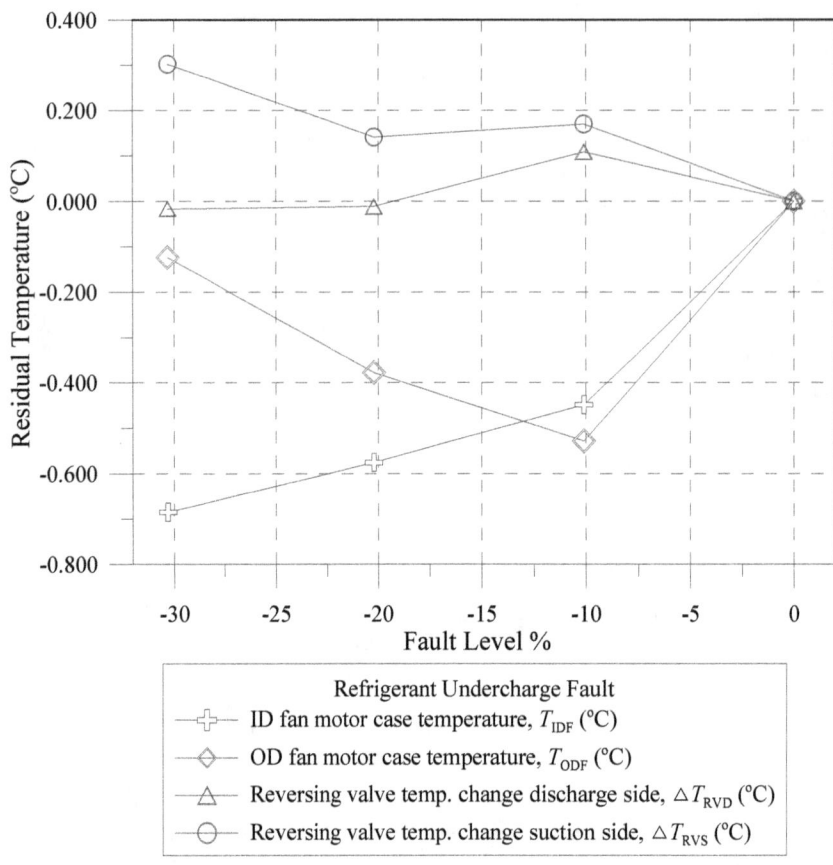

Figure 5.6.5. Residual of selected features with undercharged refrigerant faults at indoor dry-bulb temperature of 21.1 °C and outdoor conditions of -8.3 °C/Dry: R[T_{IDF}], R[T_{ODF}], R[ΔT_{RVD}], and R[ΔT_{RVS}]

Table 5.6.1. Linear fit residual slopes and feature changes as a function of percent decrease in refrigerant charge at indoor dry-bulb temperature of 21.1 °C and outdoor conditions of -8.3 °C/Dry

Feature Name	Feature Symbol	Feature's slope	Feature's % Change @ Max Fault Level
total air side capacity (kW %$^{-1}$)	Q_{CA}	8.04E-03	-4.9
compressor power (kW %$^{-1}$)	W_{comp}	1.55E-03	-2.9
(INVALID) refrigerant mass flow rate (kg min^{-1} %$^{-1}$)	m_R	7.93E-03	-17.3
coefficient of performance, air-side (%$^{-1}$)	COP	2.00E-03	-2.8
(INVALID) indoor unit refrigerant side capacity (kW %$^{-1}$)	Q_{CR}	2.89E-02	-17.6

Refrigerant mass was removed. Fault determined by % decrease below normal charge level determined during the cooling mode.	Max Fault Level: -30.3 %

Listed in Descending Order of Largest ABS(ΔT°C (% Fault)$^{-1}$)

		Residual's slope as a function of % fault level	
		Δ°C (% Fault)$^{-1}$	Feature % Change @ Max Fault Level
liquid line subcooling at outdoor service valve (°C)	ΔT_{scV}	0.196	-89.7
liquid line temperature drop (°C)	ΔT_{LL}	0.096	-68.4
condenser inlet saturation temperature (°C)	T_{CR}	0.045	-2.8
condenser bend thermocouple, TC#15 (°C)	T_{C15}	0.040	-2.4
vapor superheat at outdoor service valve (°C)	ΔT_{shV}	-0.023	2.3
ID fan motor case temperature (°C)	T_{IDF}	0.022	-0.9
compressor discharge wall temperature (°C)	T_D	0.021	-0.9
condenser inlet superheat (°C)	ΔT_{shC}	-0.020	2.4
condenser air temperature rise (°C)	ΔT_{CA}	0.017	-5.7
evaporator exit superheat (°C)	ΔT_{shE}	-0.016	8.8
reversing valve temperature change, suction side (°C)	ΔT_{RVS}	-0.009	14.5
evaporator exit saturation temperature (°C)	T_{ER}	0.008	-11.5
outdoor temperature minus TC#103 (°C)	ΔT_{103}	-0.007	5.0
evaporator air temperature drop (°C)	ΔT_{EA}	0.007	-8.3
evaporator bend thermocouple, TC#103 (°C)	T_{E103}	0.004	-2.9
OD fan motor case temperature (°C)	T_{ODF}	0.002	-0.3
reversing valve temperature change, discharge side (°C)	ΔT_{RVD}	0.002	-0.9

5.6.2   Indoor Dry-Bulb Temperature of 21.1 °C at Outdoor Conditions of 8.3 °C/Dry

Figure 5.6.6 shows the change in air-side heating capacity, refrigerant-side heating capacity, compressor power, COP, and refrigerant mass flow rate as a function of the percent decrease in refrigerant charge (% decrease is a negative number). Air-side heating capacity and air-side COP changed by -13.9 % and -9.2 %, respectively, with a -30.3 % change in refrigerant mass charged. Normally, refrigerant and air side measurements would be within ±3 % or better, but refrigerant mass flow rate, and thus refrigerant-side capacity, was affected by the presence of bubbles in the liquid line.

Figure 5.6.7 shows the residuals of T_{ER}, T_{E103}, and ΔT_{shE} as a function of refrigerant undercharge. The evaporator exit superheat, ΔT_{shE}, remained stable until a more than 20 % reduction in charge due to corrective actions by the thermostatic expansion valve. At the maximum fault level ΔT_{shE} increased to be 2.4 °C greater than its NFSS value. The pressure calculated value of the evaporator exit saturation temperature, T_{ER}, and the thermocouple measured value, T_{E103}, mirrored the changes in superheat residual.

Figure 5.6.8 shows the residuals of T_D, T_{CR}, T_{C15}, ΔT_{shC}, ΔT_{shV}, and ΔT_{scV}. Liquid line subcooling at the service valve (ΔT_{scV}), had the greatest change due to an undercharge fault. At the maximum fault level of -30.3 %, the liquid line subcooling droped by 3.5 °C from its NFSS value. The residuals of the condenser inlet saturation temperature and its thermocouple counterpart, T_{CR} and T_{C15}, also showed large slopes.

Figure 5.6.9 shows the residuals of ΔT_{CA}, ΔT_{EA}, ΔT_{LL}, and ΔT_{103}. This figure shows that the liquid line temperature change was not as large as it was at the lower outdoor temperature. The indoor coil air temperature rise residual showed the most change of all these features. The outdoor air temperature minus the bend thermocouple, TC#103, showed the second largest residual slope for these features; ΔT_{103} had a residual slope of -0.048 C %$^{-1}$ and changed by more than 25 % at the maximum fault level.

Figure 5.6.10 shows the residuals of T_{IDF}, T_{ODF}, ΔT_{RVD}, and ΔT_{RVS}. The indoor fan case temperature residual showed the strongest correlation to refrigerant undercharge. The second largest residual slope occurred for the outdoor fan motor case temperature.

Table 5.6.2 shows the residual's linear slopes and the percent changes in the system characteristics and temperatures for the undercharged refrigerant fault. The prominent features for the undercharge fault are liquid line subcooling, condenser inlet saturation temperature, and vapor line superheat at the service valve. Interestingly, the indoor fan case temperature residual was comparable to the service valve superheat residual for indicating an undercharge fault.

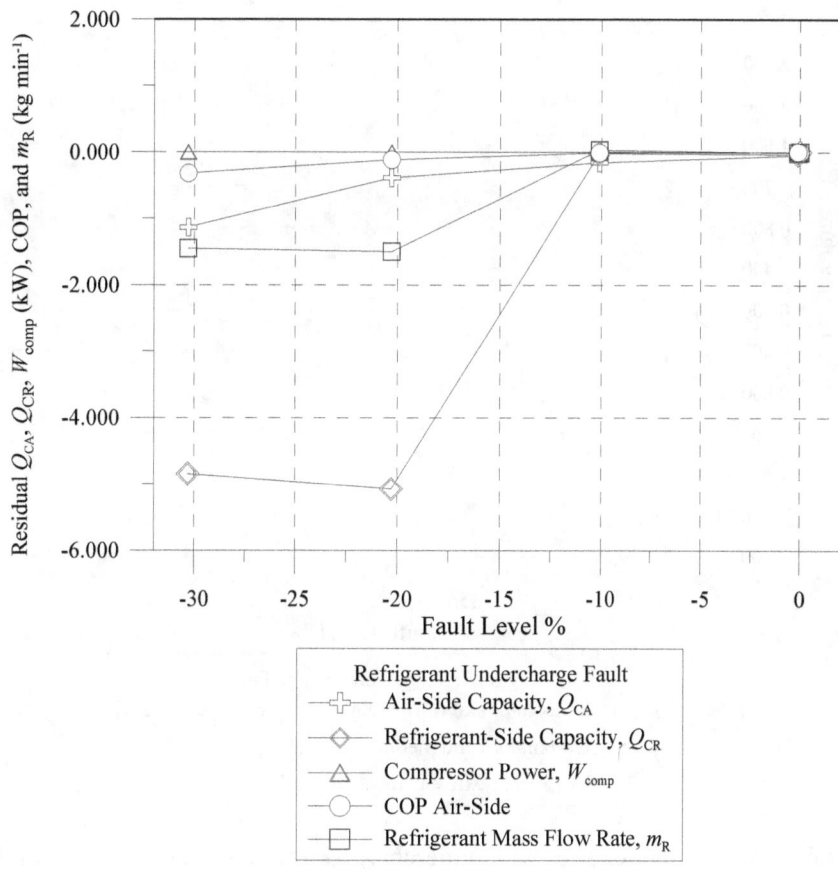

Figure 5.6.6. Residual of selected features with undercharged refrigerant faults imposed at an indoor dry-bulb temperature of 21.1 °C and outdoor conditions of 8.3 °C/Dry: R[Q_{CA}], R[Q_{CR}], R[W_{comp}], R[COP], and R[m_R]

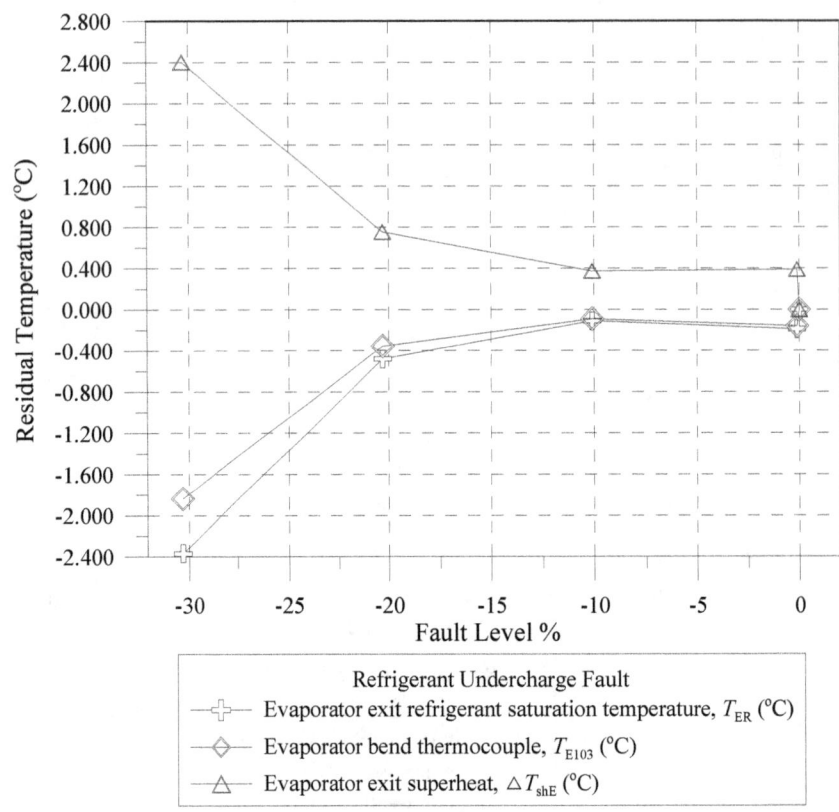

Figure 5.6.7. Residual of selected features with undercharged refrigerant faults at indoor dry-bulb temperature of 21.1 °C and outdoor conditions of 8.3 °C/Dry: $R[T_{ER}]$, $R[T_{E103}]$, and $R[\Delta T_{shE}]$

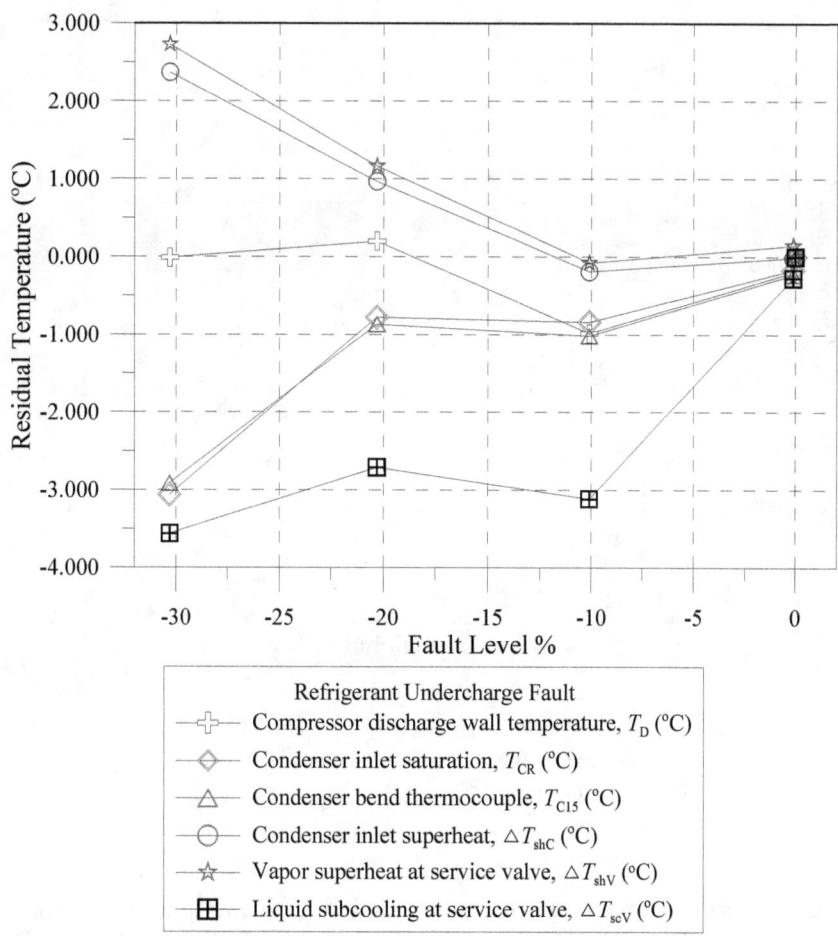

Figure 5.6.8. Residual of selected features with undercharged refrigerant faults at indoor dry-bulb temperature of 21.1 °C and outdoor conditions of 8.3 °C/Dry: R[T_D], R[T_{CR}], R[T_{C15}], R[ΔT_{shC}], R[ΔT_{shV}], and R[ΔT_{scV}]

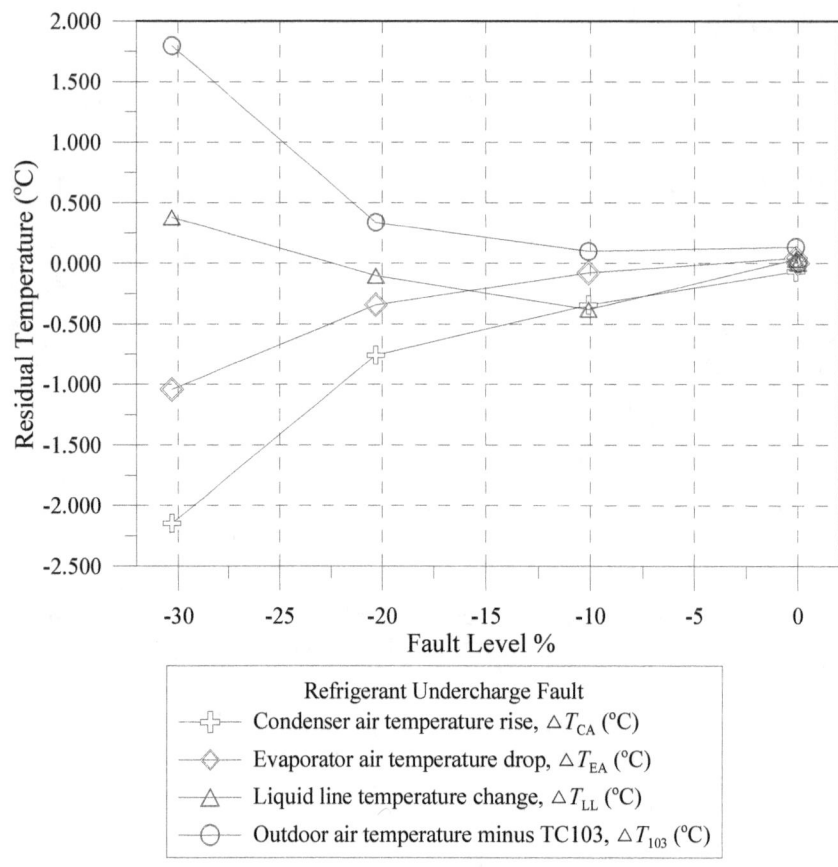

Figure 5.6.9. Residual of selected features with undercharged refrigerant faults at indoor dry-bulb temperature of 21.1 °C and outdoor conditions of 8.3 °C/Dry: R[ΔT_{CA}], R[ΔT_{EA}], R[ΔT_{LL}], and R[ΔT_{103}]

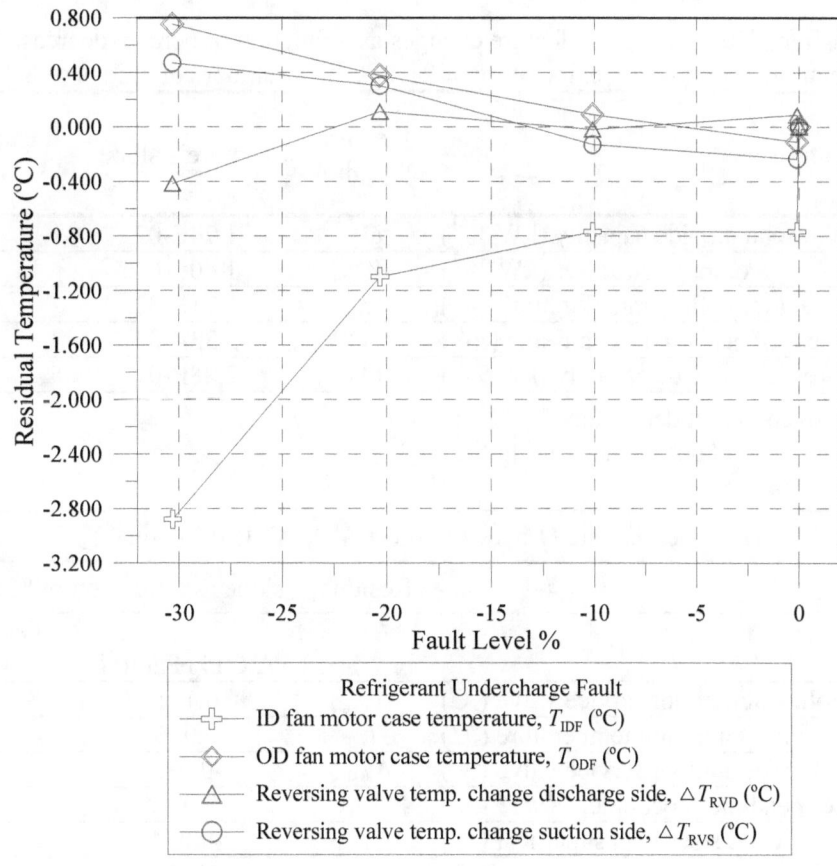

Figure 5.6.10. Residual of selected features with undercharged refrigerant faults at indoor dry-bulb temperature of 21.1 °C and outdoor conditions of 8.3 °C/Dry: R[T_{IDF}], R[T_{ODF}], R[ΔT_{RVD}], and R[ΔT_{RVS}]

Table 5.6.2. Linear fit residual slopes and feature changes as a function of percent decrease in refrigerant charge at indoor dry-bulb temperature of 21.1 °C and outdoor conditions of 8.3 °C/Dry

Feature Name	Feature Symbol	Feature's slope	Feature's % Change @ Max Fault Level
total air side capacity (kW %$^{-1}$)	Q_{CA}	1.74E-02	-13.9
compressor power (kW %$^{-1}$)	W_{comp}	1.40E-03	-6.6
(INVALID) refrigerant mass flow rate (kg min^{-1} %$^{-1}$)	m_R	6.71E-02	-61.1
coefficient of performance, air-side (%$^{-1}$)	COP	5.33E-03	-58.1
(INVALID) indoor unit refrigerant side capacity (kW %$^{-1}$)	Q_{CR}	2.28E-01	-60.3

Refrigerant mass was removed. Fault determined by % decrease below normal charge level determined during the cooling mode.	Max Fault Level: -30.3 %

Listed in Descending Order of Largest ABS(ΔT°C (% Fault)$^{-1}$)

		Residual's slope as a function of % fault level	
		Δ°C (% Fault)$^{-1}$	Feature % Change @ Max Fault Level
liquid line subcooling at outdoor service valve (°C)	ΔT_{scV}	0.111	-87.7
condenser inlet saturation temperature (°C)	T_{CR}	0.084	-5.5
vapor superheat at outdoor service valve (°C)	ΔT_{shV}	-0.083	10.5
condenser bend thermocouple, TC#15 (°C)	T_{C15}	0.079	-5.1
condenser inlet superheat (°C)	ΔT_{shC}	-0.074	10.0
ID fan motor case temperature (°C)	T_{IDF}	0.073	-3.4
evaporator exit saturation temperature (°C)	T_{ER}	0.064	-13.7
evaporator exit superheat (°C)	ΔT_{shE}	-0.064	35.9
condenser air temperature rise (°C)	ΔT_{CA}	0.063	-14.4
evaporator bend thermocouple, TC#103 (°C)	T_{E103}	0.049	-9.6
outdoor temperature minus TC#103 (°C)	ΔT_{103}	-0.048	25.7
evaporator air temperature drop (°C)	ΔT_{EA}	0.032	-21.0
OD fan motor case temperature (°C)	T_{ODF}	-0.026	1.6
reversing valve temperature change, suction side (°C)	ΔT_{RVS}	-0.020	62.7
reversing valve temperature change, discharge side (°C)	ΔT_{RVD}	0.011	-12.8
liquid line temperature drop (°C)	ΔT_{LL}	-0.009	17.4
compressor discharge wall temperature (°C)	T_D	-0.008	0.0

5.7 Summary of the Effects of Faults on Heating Capacity and COP

Figure 5.7.1 shows the effects of condenser or indoor coil air flow restriction faults on the heating capacity and COP. It is interesting to note the linear nature of the normalized values with increasing air flow restriction across the indoor heat exchanger. Higher fault levels indicate more reduction in the air flow rate across the indoor coil (condenser). At higher outdoor temperature the fault had a greater effect upon capacity and COP. There was a 10 % reduction in COP with a 30 % reduction in indoor coil air flow rate at outdoor ambient conditions of 8.3 °C/72.5 % RH.

Figure 5.7.2 shows the effects of outdoor coil face area blockage on heating capacity and COP. A 30 % blockage of face area resulted in a 26 % reduction in heating capacity and a 24 % reduction in COP at the coldest outdoor ambient conditions. The same fault was not as severe for the higher ambient conditions; heating capacity dropped by 18 % while COP decreased by 14 %.

Figure 5.7.3 shows how a compressor valve or a four-way reversing valve leakage fault affects heating capacity and COP. Heating capacity at the higher temperature outdoor conditions was more affected by this fault; heating capacity decreased by almost 14 % as refrigerant mass flow rate decreased more than 12 %. COP degradation was comparable at the higher and lower outdoor ambient conditions.

Figure 5.7.4 shows that the liquid line restriction fault had very little effect upon the functioning of the system in the heating mode; the obtained results are within the uncertainty of our measurements.

Figure 5.7.5 shows that refrigerant overcharge has negligible effects on heating capacity, but more of an effect upon COP. At the higher temperature outdoor ambient conditions, the COP dropped by more than 14 % as the refrigerant was overcharged 30 %.

Figure 5.7.6 shows that undercharge faults have more effect upon heating capacity than the overcharge faults, especially at higher temperature outdoor ambient conditions. For a slightly more than 30 % undercharge fault, heating capacity decreased by almost 14 % while COP dropped by almost 9 %.

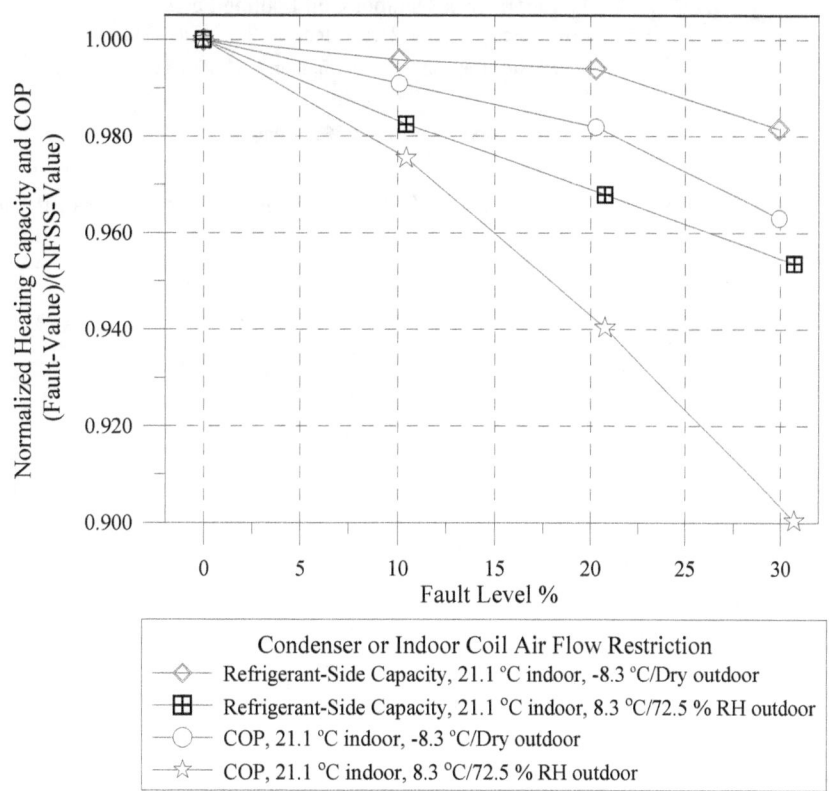

Figure 5.7.1. Normalized heating capacity and COP for condenser or indoor coil air flow restriction

151

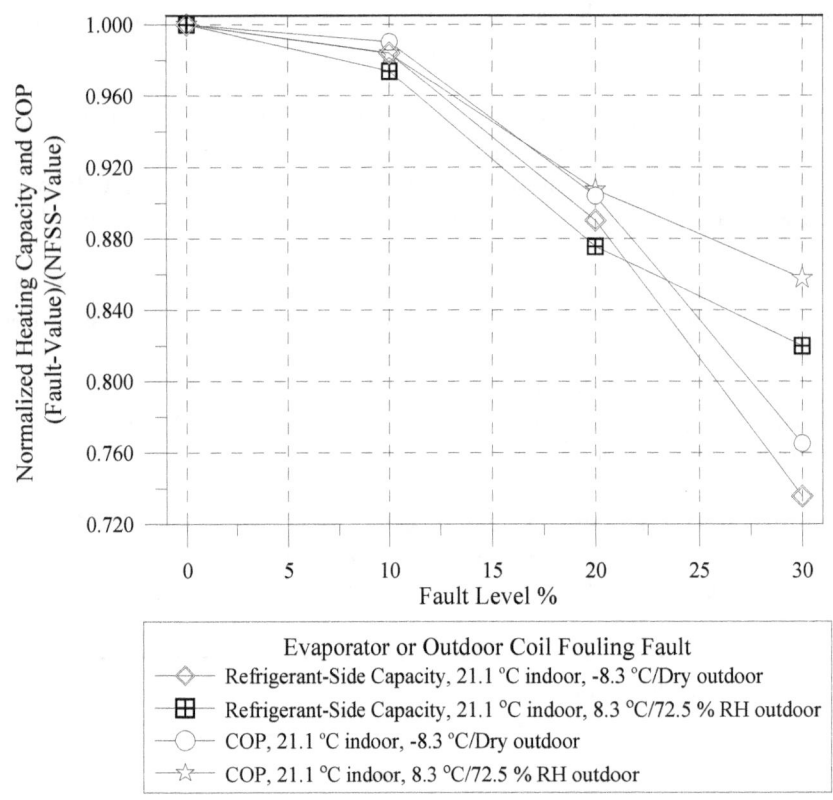

Figure 5.7.2. Normalized heating capacity and COP for evaporator or outdoor coil fouling faults

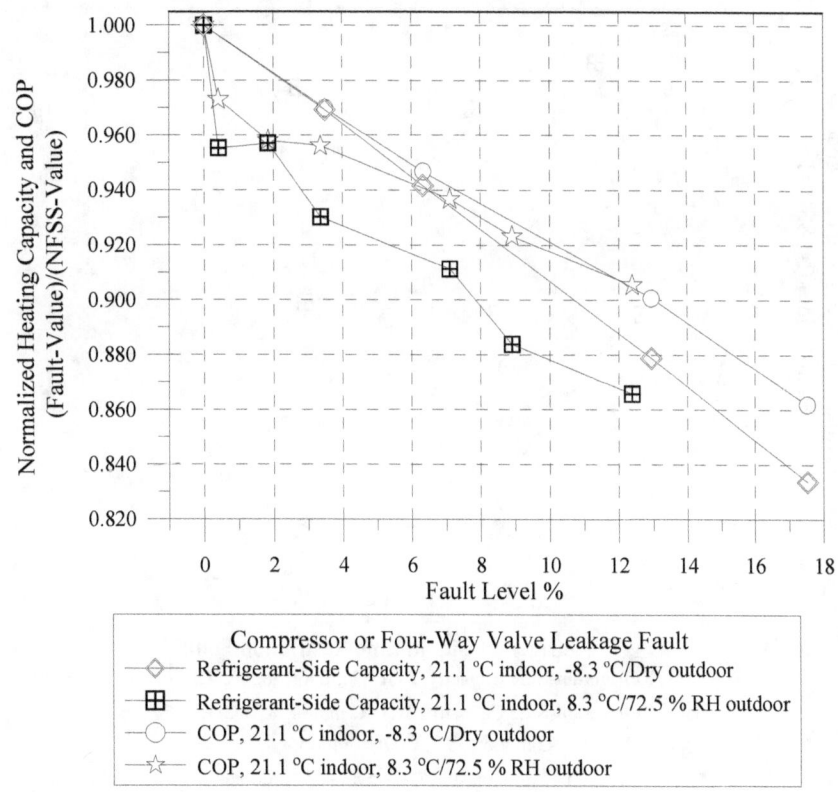

Figure 5.7.3. Normalized heating capacity and COP for compressor or four-way valve leakage faults

153

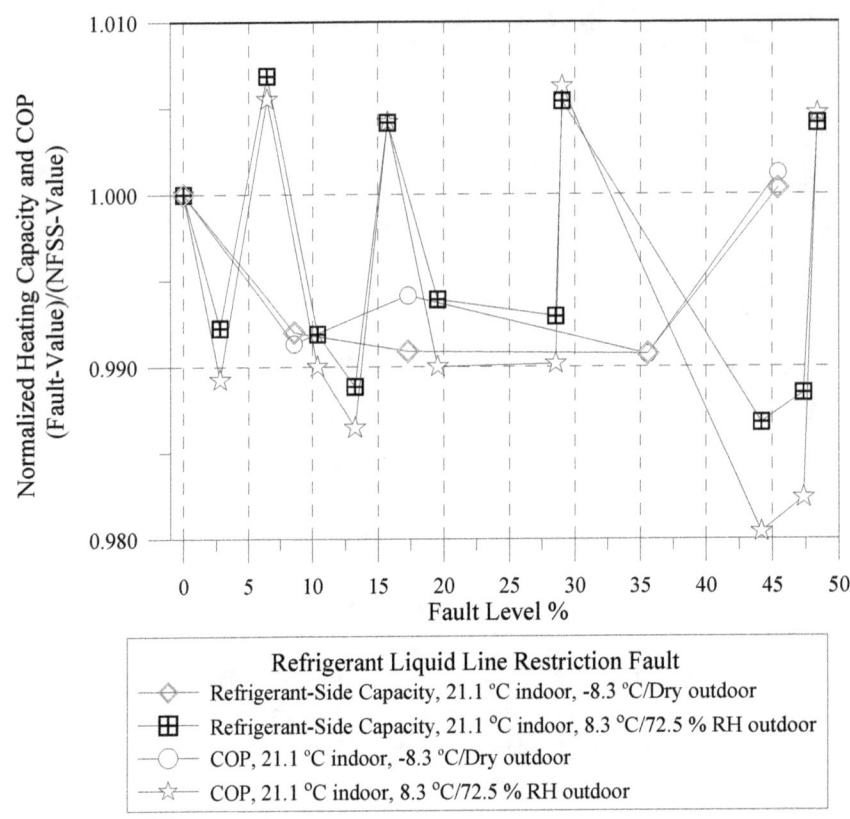

Figure 5.7.4. Normalized heating capacity and COP for refrigerant liquid line flow restriction faults

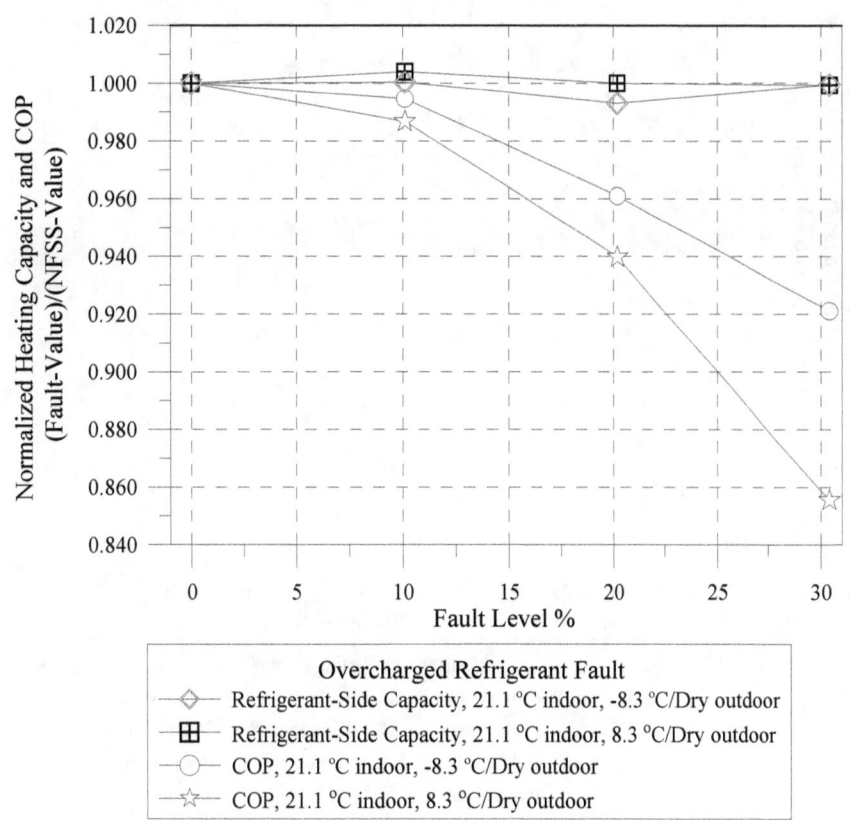

Figure 5.7.5. Normalized heating capacity and COP for refrigerant overcharge faults

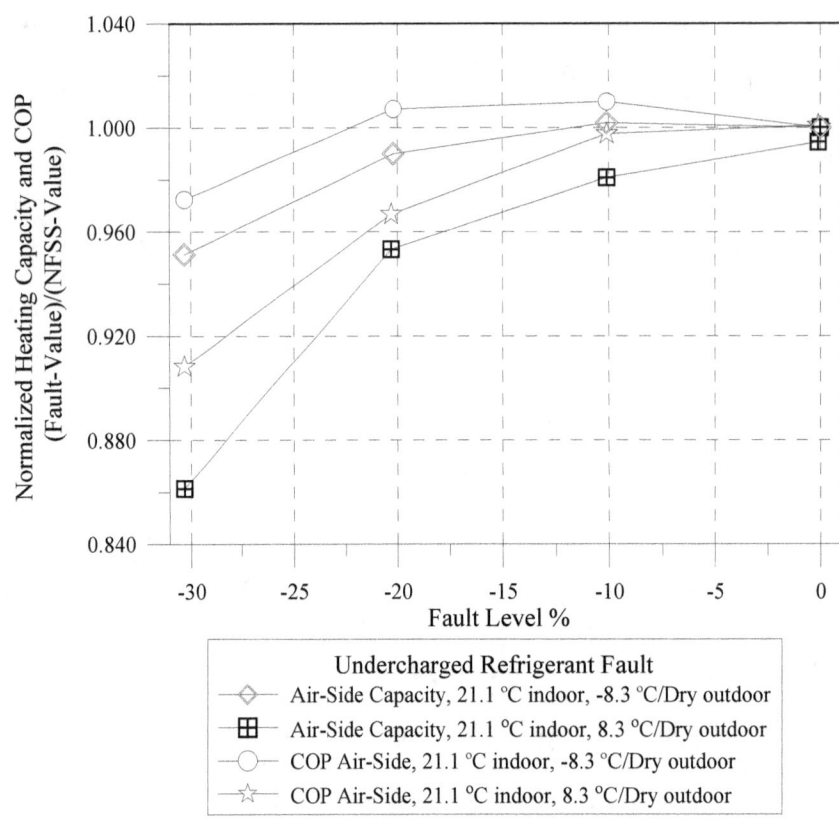

Figure 5.7.6. Normalized heating capacity and COP for refrigerant undercharge faults

CHAPTER 6. Concluding Remarks

A residential heat pump was tested in the heating mode at several indoor and outdoor conditions with no faults and then with faults imposed. The no-fault performance of the system will be used to generate a No-Fault Steady-State (NFSS) reference model of various system features as a function of the independent variables of indoor dry-bulb, outdoor dry-bulb and, possibly, outdoor dew-point temperature. Knowing NFSS values allows the calculation of feature residuals. As in the previous work by Kim et al. (2006), certain patterns of feature residuals represent certain faults. The magnitude of the feature residuals is also directly related to the probability that a fault is occurring. With the data collected in the heating mode, a complete FDD method will be developed to detect the studied faults.

Heating capacity was affected most by the compress/reversing valve leakage fault (CMF). For the constant indoor dry-bulb temperature of 21.1 °C, heating capacity decreased by 70 W per % fault at 8.3 °C outdoor temperature and 43 W per % fault at -8.3 °C outdoor temperature. COP was also most affected by the CMF fault with a 22.3e-3 decrease per % fault increase at 8.3 °C and 17.6e-3 decrease per % fault at -8.3 °C. The evaporator or outdoor coil fouling fault (EF) had the second greatest effect upon heating capacity and COP; reducing heating capacity and COP by 53.4 W per % fault and 17.5e-3 per % fault at the 8.3 °C outdoor temperature, respectively. With the exception of the liquid line restriction fault, all faults had the most affect at higher capacity, higher temperature outdoor conditions.

For the test conditions studied in this work, the liquid line restriction fault did not pose a penalty on COP or heating capacity until a fault level greater than 48 %. Other faults that required severe levels to produce a 5 % reduction in COP or heating capacity were condenser air flow fault (max 50% at low temperatures), undercharged refrigerant, and overcharged refrigerant. At an outdoor temperature of 8.3 °C, the refrigerant had to be more than 18 % overcharged to produce a 5 % drop in COP, and , if we extrapolate the trend, more than 50 % overcharged to produce the same loss in heating capacity. Undercharge produced a 5 % drop in COP and heating capacity at a fault level of approximately 25 %.

Outdoor coil frosting also poses a unique problem for heating mode fault detection. As seen in the figures of chapter 4, system features change substantially as the outdoor coil frosts; some features change by more than 50 % during frosting. A reliable method of frost detection must be implemented within any heating mode FDD algorithm. The fault detection method could rely on temperature and humidity sensors at the outdoor unit, monitoring of various features that change substantially with frosting, or a combination of these methods to determine when frosting is occurring. Frosting density and frosting rates vary too much to be included in an FDD algorithm, thus frosting must be treated as a type of fault that the FDD algorithm tries to detect. Frosting conditions can simply be detected by comparing an evaporator (outdoor coil) two-phase return bend temperature to some temperature above freezing; FDD would only be implemented outside of frosting conditions.

The results presented here show the difficulty in detecting faults in the heating mode. The work by Li et al. (2007) in developing virtual sensors and in isolating faults will aid in developing generalized FDD techniques. The uniqueness of each heat pump installation will make a robust and sensitive FDD system very difficult to develop without some kind of "machine learning" or adaptive correlation technique to adjust the FDD algorithms to each different type of installation environment seen in the field. As seen in the previous work by Kim et al. (2006) and the work presented here, most of the features change in a relatively linear manner as fault levels increase; this should make it easier to implement adaptive correlations for predicting fault-free feature values on small microprocessors within the system's normal controls. A simple FDD algorithm could easily monitor system refrigerant charge using only two temperature sensors (refrigerant subcooling for a TXV system); two more sensors could be added to monitor refrigerant superheat and thus become the basis for a simple commissioning tool to aid technicians in setting the proper charge during the intial installation. The common problem of charging a system in the heating mode could also be aided by an FDD system that monitors the compressor discharge temperature; compressor refrigerant discharge temperature was the most sensitive variable to refrigerant overcharge in the heating mode tests performed for this work.

REFERENCES

ANSI/ACCA Standard 5, 2007, HVAC quality installation specification, Air Conditioning Contractors of America (www.acca.org), 2800 Shirlington Road, Suite 300, Arlington, VA 22206.

Anderson, D., Graves, L., Reinert, W., Kreider, J.F., Dow, J., and Wubbena, H., 1989, "A Quasi-Real-Time Expert System for Commercial Building HVAC Diagnostics," *ASHRAE Transactions*, Vol. 95, Part 2, pp. 25-28.

ARI Standard 210/240, 2006, Standard for unitary air-conditioning and air-source heat pump equipment, Air-Conditioning and Refrigeration Institute, 4100 North Fairfax Drive, Suite 200, Arlington, VA 22203.

Braun JE. 1999, Automated fault detections and diagnostics for vapor compression cooling equipment. Int J HVAC&R Research, 5(2);85-86.

Breuker, M.S. and Braun, J.E., 1998, "Common faults and their impacts for rooftop air conditioners," *International Journal of Heating, Ventilating, Air-Conditioning and Refrigerating Research*, 4(3), 303-18.

Breuker, M.S., Rossi, T.M. and Braun, J.E., 2000, "Smart maintenance for rooftop units," *ASHRAE Journal*, 42(11), 41-6.

Brownell, K.A., Reindl, D.T. and Klein, S.A., 1999, "Refrigeration system malfunctions," *ASHRAE Journal*, 41(2), 40-7.

Castro, N.S., 2002, "Performance Evaluation of a Reciprocating Chiller Using Experimental Data and Model Predictions for Fault Detection and Diagnosis," *ASHRAE Transactions*, 108(1), 889-903.

Chen, B. and Braun, J.E., 2001, "Simple Rule-Based Methods for Fault Detection and Diagnostics Applied to Packaged Air Conditioners," *ASHRAE Transactions*, 107, (1), 847-857.

Comstock, M.C., Braun, J.E., and Groll, E.A., 2001, "The Sensitivity of Chiller Performance to Common Faults," *HVAC&R Research*, Vol. 7, No. 3, pp. 263-279.

Grimmelius, H.T., Woud, J.K., and Been, G., 1995, "On-line Failure Diagnosis for Compression Refrigeration Plants," *International Journal of Refrigeration*, Vol. 18, No. 1, pp. 31-41.

Hayter, S.J., Torcellini, P.A. and Judkoff, R. 1999, "Optimizing building and HVAC systems," *ASHRAE Journal*, 41(12), 46-49.

Kim, M. and Kim, M.S., 2005, "Performance Investigation of a Variable Speed Vapor Compression System for Fault Detection and Diagnosis," *International Journal of Refrigeration*, 28(4), 481-88.

Kim, M., Payne, W.V., Hermes, C.J.L., and Domanski, P.A., 2006, "Performance of a Residential Air Conditioner at Single-Fault and Multiple-Fault Conditions," NISTIR 7350, National Institute of Standards and Technology, Gaithersburg, MD.

Kim, M., Payne, W.V., Yoon SH, and Domanski, P.A., 2008a, "Cooling Mode Fault Detection and Diagnosis Method for a Residential Heat Pump," NIST Special Publication 1087, National Institute of Standards and Technology, Gaithersburg, MD.

Kim, M., Yoon, S. H., Payne, W. V., and Domanski, P. A., 2008b, "Design of a steady-state detector for fault detection and diagnosis of a residential air conditioner," International Journal of Refrigeration 31(5), 790-99.

Lee, W.Y., Park, C., and Kelly, G.E., 1996a, "Fault Detection in an Air-Handling Unit Using Residual and Recursive Parameter Identification Methods," *ASHRAE Transactions*, 102(1), 528-39.

Lee, W.Y., House, J.M., Park, C., and Kelly, G.E., 1996b, "Fault Diagnosis of Air-Handling Unit Using Artificial Neural Networks," *ASHRAE Transactions*, 102(1), 540-49.

Li H. 2004, A decoupling-based unified fault detection and diagnosis approach for packaged air conditioners. Ph.D. Thesis, West Lafayette, IN: Purdue University.

Li, H. and Braun, J.E., 2003, "An Improved Method for Fault Detection and Diagnosis Applied to Packaged Air Conditioners," *ASHRAE Transactions*, 109(2), 683-92.

Li, H. and Braun, J.E., 2007. Decoupling features and virtual sensors for diagnosis of faults in vapor compression air conditioners, *International Journal of Refrigeration*, 30, 546-64.

Li, H. and Braun, J.E., 2009. Decoupling features for diagnosis of reversing and check valve faults in heat pumps, *International Journal of Refrigeration*, 32(2), 316-26.

McKellar, M.G., 1987. "Failure Diagnosis for a Household Refrigerator," M.S. Thesis, Purdue University, West Lafayette, IN.

Norford, L.K. and Little, R.D., 1993, "Fault detectionand load monitoring in ventilation systems," *ASHRAE Transactions*, 99(1), 590-602.

Payne, W.V., Domanski, P.A., and Muller, J., 1999, "A Study of a Water-to-Water Heat Pump using Hydorcarbon and Hydrofluorocarbon Zeotropic Mixtures," NISTIR 6330, National Institute of Standards and Technology, Gaithersburg, MD.

Pape, F.L.F., Mitchell, J.W., and Beckman, W.A., 1991, "Optimal Control and Fault Detection in Heating, Ventilating, and Air-Conditioning System," *ASHRAE Transactions*, Vol. 97, Part 1, pp. 729-745.

Peitsman, H.C. and Bakker, V.E., 1996, "Application of Black-Box Models to HVAC Systems for Fault Detection," *ASHRAE Transactions*, Vol. 102, Part. 1, pp. 628-640.

Proctor, J., 2002, "What can 13,000 air conditioners tell us?" American Council for an Energy Efficient Economy (ACEEE), Summer study on energy efficiency in buildings, 12th bienial building summer study, Paper 431, Session 1, pages 53-68.

Proctor, J., 2004, "Residential and Small Commercial Central Air Conditioning; Rated Efficiency isn't Automatic," Presentation at the Public Session. ASHRAE Winter Meeting, January 26, Anaheim, CA.

Rossi, T.M. and Braun, J.E., 1997, "A statistical, rule-based fault detection and diagnostic method for vapor compression air conditioners, *International Journal of Heating, Ventilating, Air-Conditioning and Refrigerating Research*," 3(1), 19-37.

Rossi, T.M., 2004, "Unitary Air Conditioner Field Performance," International Refrigeration and Air Conditioning Conference at Purdue, Paper No. R146, July 12-15, West Lafayette, IN.

Roth, K., Llana, P., Westphalen, D., and Brodrick, J., 2005, "Emerging technologies: automated whole building diagnostics," *ASHRAE Journal*, 47(5), 82-84.

Seem, J.E., House, J.M. and Monroe, R.H., 1999, "On-line monitoring and fault detection," *ASHRAE Journal*, 41(77), 21-26.

Smith, V.A. and Braun, J.E., 2003, "Fault Detection and Diagnostics for Rooftop Air Conditioners," Final Report Compilation for Project 2.1, Publication #P500-03-096, California Energy Commission, http://www.archenergy.com/cec-eeb/reports.htm

Snoonian, D., 2003, "Smart buildings," IEEE Journal, 40(8), 18-23.

Stylianou, M. and Nikanpour, D., 1996, "Performance Monitoring, Fault Detection and Diagnosis of Reciprocating Chillers," *ASHRAE Transactions*, 102(1), 615-27.

Stylianou, M., 1997, "Application of Classification Functions to Chiller Fault Detection and Diagnosis," *ASHRAE Transactions*, 103(1), 645-56.

Westphalen, D., Roth, K.W., and Brodrick, J., 2003, "System and Component Diagnostics," *ASHRAE Journal*, Vol. 45, No. 4, pp. 58-59.

www.ingramcontent.com/pod-product-compliance
Lightning Source LLC
Chambersburg PA
CBHW080246180526
45167CB00006B/2432